Decentering Science and Technology in Global Development

Integrated Global STEM

Edited by
Robert Krueger, Wole Soboyejo and Romain Murenzi

Volume 6

Decentering Science and Technology in Global Development

Co-Designing for Community Sustainability

Edited by
Robert Krueger

DE GRUYTER The WPI Press

Editor
Prof. Robert Krueger
Worcester Polytechnic Institute
Institute of Science and Technology for Development
100 Institute Road
Worcester, MA 01609
USA
krueger@wpi.edu

ISBN 978-3-11-914352-3
e-ISBN (PDF) 978-3-11-221871-6
e-ISBN (EPUB) 978-3-11-221910-2

Library of Congress Control Number: 2025942290

Bibliographic information published by the Deutsche Nationalbibliothek
The Deutsche Nationalbibliothek lists this publication in the Deutsche Nationalbibliografie; detailed bibliographic data are available on the internet at http://dnb.dnb.de.

Cover image: Robert Krueger
Typesetting: Integra Software Services Pvt. Ltd.

www.degruyterbrill.com
Questions about General Product Safety Regulation:
productsafety@degruyterbrill.com

Preface

This is a timely volume. Between the time I formally submitted this manuscript in mid-January 2025, I breathed another sigh of relief to have this thing out of my hands for a minute. The week after I submitted the manuscript the momentum began to end USAID not just as we knew it, but quite literally to end it. An executive order was passed on January 20, pausing US-funded development projects for 90 days. I gave this little thought. We'd been to this dance before. No one else I spoke to believed it either. It's too big. It means too much to the US national security and soft foreign relations. Then, on Friday, January 24, the unthinkable happened, stop work orders were emailed, the famous fork in the road email was published, and a week or so later people were locked out of their offices. The world of international humanitarian and development continues its state of upheaval. As usual, the United States came late to the party (the UK, Sweden, Belgium, Germany, and France have cut back aid significantly in recent years). When America did show up – as usual – it changed from a gathering into something like *Burning Man* during the rains.

I am sad for my many friends who worked at the U.S. Agency for International Development (USAID) and their implementing partners. The Administration says it cut 10,000 federal jobs by integrating USAID into the State Department. What is not mentioned is that 100,000 other people lost their jobs when USAID folded. Some put that number at around 175,000 jobs lost and counting. My friends not only lost their jobs but also lost the entire economic sector they worked in; there were no other related jobs to go to. Many are still reeling from the shock and despair of losing not only their job, but a mission that they believed in completely. They sacrificed sleep, being at home with, and endured long temporary duty assignments. In return, they got malaria and the runs, but they also got satisfaction from their work, authentic relationships with local collaborators, and the different ways of doing that they encountered. For those of us who have had them, these experiences are priceless. They truly decenter Western ways of doing from 'the right way' to 'one way' of doing things.

A couple of months later, I began to hear different perspectives. For example, at a World Economic Forum–sponsored event called the Global AI Summit on Africa, I heard from many African leaders who welcomed the idea of a post-USAID (post-Western) approach to development. These leaders seemed ready, if not eager, to witness the end of Western hegemony and seize the opportunity for self-guided development. To understand what could happen one only has to look back to a newly independent Ghana in the late 1950s when self-determination, the legacy of colonialism, and the politics of Western domination collided only to reproduce the worst of Western ideology and practice (see Chapter 1 for more details).

These events and perspectives make a volume like this even more important that it was before January 20, 2025. In the first iteration of this preface, I wrote:

https://doi.org/10.1515/9783112218716-202

> Perhaps it's not a coincidence that a volume like this one has had such a difficult time coming to your digital or physical library. This book, like the topics it addresses, has been fraught with unforeseen setbacks and unanticipated consequences. The emergence of this book has mirrored in some ways the process we have called for in the following pages for engineering for development projects.

The audience of this book was, and remains, the people who design, fabricate, construct, evaluate, and innovate in the development space. While this book is critical of development as practiced, students of anthropology, geography, or development studies will, out the outset, not find what tied in the bow of French post-structuralist philosophy. Our purpose is not to break new ground on theories of dependency, post-development, or decoloniality.

Rather, our primary purpose is to speak to those technicians responsible for designing and implementing interventions so that they move beyond *solutionism* – the belief that all complex social problems can be solved with technology. More than this, the volume offers the reader perspectives and practices for moving beyond solutionism, instrumentalism, and cultural centrism. For the development studies crowd, broadly construed, the book brings insight into the thinking of scientists and engineers who engage in development. This is a *text* open for interpretation.

One might think that it guileless to produce a volume like this with contributors from – primarily – one institution, Worcester Polytechnic Institute (WPI). Because we spend our careers communicating our disciplinary ideas to each other and to our students, I would say that there is no better place from such a volume to emerge. The contributors to this volume, whether a roboticist, a civil engineer, an anthropologist, or a geographer, are spending their careers in teaching and learning with students and colleagues from a STEM forward university. This perspective provides us a unique opportunity to take the critique, pare it down to its most basic terms, and make it digestible for those who might think Foucault rhymes with default. For our STEM colleagues, they, too, do a great deal of translating and communicating the ideas in this book from their perspective as engineers. The capacity to develop fluency in different disciplines is the basis of transdisciplinary practice (see Chapter 3). Suffice it to say, our shared goal is to learn from one another, digest their perspective, and create a new synthesis that neither would have thought of before any conversation began.

This intellectual elasticity is made possible by WPI's relatively modest size and *the WPI Plan*. The WPI Plan was created over 50 years ago when a generation of our faculty realized that STEM education for its own sake was an incomplete education. Their plan did away with majors, grades, leaned into a barely known concept called project-based learning, and they developed humanities and social science departments to broaden a student's perspective. Indeed, both of these latter departments have responsibility for core WPI degree requirements. In addition, WPI sends 1,200 students and hundreds of faculty abroad annually to work on community-based projects, on 6 continents, that require students and faculty to interrogate technical questions and interventions through social and cultural lenses. These projects are embedded in places ranging from informal settlements, like Katatura, Namibia, the riverside slums of Bangkok, the traditional lands of Akyem Abuakwa in Ghana, to Cuenca, Ecuador, and San Jose, Costa Rica; and, finally, our own multicultural community of Worcester, MA. WPI is a global institution that is engaged across many global communities. Through these engagements, we bring our expertise and learn for others while we facilitate problem-solving through co-creation informed by generative justice.

Robert Krueger, Worcester, MA

Contents

Part I

Robert Krueger

Chapter 1
Introduction: The Uneasy Entanglement of Society with Science and Technology in International Development

Introduction

This book is about rethinking how science and technology engage in the conceptualization of development interventions through development practice. As the title of this chapter states, we are concerned about and seek to disentangle the cocktail of science and technology and society away from their standardized western, enlightenment-based, and modernist proxies. Rather, we seek to expose the myths of the past that continue to shape how science and engineering, through technological interventions, are practiced.

Given the upheaval in the development community, the time for these ideas has come. For the vacuum created here gives a new cast of implementers a chance to do things better, and maybe even right some wrongs of 75 years of the international development paradigm. As we will see below, a change of leadership does not necessarily lead to changes in outcome. Kwame Nkrumah's Ghana is an exemplary example of this (see later in this chapter).

We believe that engineering and scientific interventions, especially Western interventions, in societies in the majority world, can no longer be left to the scientists and engineers coming from the west or channeling western ideals. In addressing these concerns, this volume brings new thinking, new concepts, and processes and procedures from a variety of disciplines and analytical lenses. In the chapters that follow, the authors, some of whom are scientists and engineers, but also anthropologists, geographers, and political scientists disentangle the sociotechnical relations so that we can better understand the challenges that have befallen the global majority. We also integrate educational examples for how to train STEM forward professionals on how they can be agents of change, while maintaining a certain level of humility. The chapters that follow offer critical and compelling insights into what is wrong with current hegemonic paradigms of development and the concepts and tools we can use to advance a different direction that promotes dignity, community self-sufficiency, and sustainability.

To begin this journey, this chapter lays out what we think are some of the most significant misunderstandings between people who lend their expertise to understand community problems and ways of formulating solutions. In doing so, we lurch away from solutionism (build it and they will come) to concepts, processes, and practices

https://doi.org/10.1515/9783112218716-001

that elucidate how communities overcome bad science, exploitative histories, centuries of colonialism, and greed to leverage their endogeny in ways that set their paths.

Before moving on to the chapters we want to take a deeper dive into three concepts that our contributors address throughout these pages: misguided cultural binaries, western hegemony, the role of actors and their perceived expertise, and the entanglement of social groups.

'Africa': 'The Early Years'

Decades before Portuguese explorer Bartolomeu Dias, in 1488, made the left turn at the southern tip of what was then called 'the Dark Continent' Europeans have pointed their gaze toward this landmass, now called Africa. In the fifteenth century the Portuguese set up outposts along the eastern coastline from modern-day Dakar, Senegal, Cape Coast, Ghana, Luanda, Angola, and Cape Town, South Africa. A decade after Dias, another Portuguese explorer, Vasco de Gama, established the first trade route to India. For centuries that followed, European place names emerged along the coast of West Africa, El Mina (the Mine in Portuguese), the Kongo, the Cape of Good Hope, and so on.

Despite their focus on the coastal settlements, by 1600 the African continent had been entirely transformed by European traders. Marx would refer this as viewing Africa as "second nature" or a place that had already fallen under a capitalist gaze (cf. Smith 1984). It was a place of abundant resources and 'primitive' cultures, religions, and people who were important, but ultimately expendable, when securing the riches the land had to offer. Europeans traded with the continent's inhabitants in commodities ranging from salt and gold to human beings. The latter being a long-standing custom between warring tribes in Africa. This arrangement occurred willy-nilly for generations. Toward the end of the nineteenth century, European powers became serious about distributing Africa's range of resources among themselves. The Conference of Berlin in 1884 marked a continent-wide agreement about how the continent (sic) would be allocated to Europe. And so, the scramble for Africa began.

A book on decentering science and technology in global development is to create an entry point for examining, and ultimately decentering, the history of Western relations with Africa. In the now famous 1899 serial from Blackwood's Magazine, Joseph Conrad (1899) offers an entry point into the human condition of King Leopold II's Congo Free State and a westerner's experience of it. Literary critics and historians have explored the text in closely such as Chinua Achebe's 1974 diatribe. For us, what is curious about the book is Marlow's account of his travels up the Congo River. Marlow is a pedant and documented his observations and experience in great detail. The protagonist's voice offers a detailed account of Kurtz's interest in rivets – not just their use in boat construction but he creates detailed image of the rivets themselves, their uses, their ubiquity, and their scarcity. Yet when Marlow endeavors to explain the details of the lives of Indigenous people living on the river and in the territory more broadly. Marlow's accounts of human suffering of Endogenous people working for their white captors, his accounts are feckless. In one breath, Marlow offers unnecessary details about rivets, the cottage of a manager, the operation of a steam engine, but cannot muster understanding of human suffering or even the ecological context where he is. This could reflect Conrad's own ignorance. Another interpretation is that Conrad is establishing the limits his protagonists understanding of a world he has no familiarity. Despite being a rational, white male, of science he fails to 'see' to 'comprehend' the unfamiliar.

Though different, these perspectives continue to this day. We seek to create markets where they 'do not' exist or rather, we do not see them. Africans have had rational economies for millennia but even the most thoughtful economists of yesterday, Polyani (1964) cannot escape the mind shackles of western thinking when he describes the Dahomey slave trade. To her credit, the self-proclaimed Economic Anthropologist, Poly Hill, spent years with Ghanaian cocoa farmers to understand their 'invisible' economies through their notions of land title, cooperation, and market exchange (Hill 1964; see also Carr 2014). Interventions base that these western notions of incentives, individualism, exchange value, and mass production are wholly inadequate as they run against the grain of the heterogeneous cultures of Africa. At their core, these interventions support the West's hunger for energy, transport, and assembly infrastructure, to name a few. We fool ourselves into believing that bringing mass production to African communities provides local benefits without realizing what local benefits already exist and how they are derived. Again, we know how to make rivets, high-entropic energy engines, and how to command the waterways. Yet, we are modern-day observers who cannot see.

The Exotic Orient?

Let's now explore this through the lens of Bruno Latour's thesis that 'we have never been modern'. Bruno Latour (1991) proposed the thought exercise that explored Western Science as an article of faith – a social construct; it is grounded in culture and politics as much as it is science. In raising this point, Latour examines what to many is sacrilege. For Latour, then, Western Science is one of potentially many cultures of inquiry.

This idea is simply demonstrated in one of his earlier works with British sociologist Steve Woolgar. In a 2-year study at the Salk Institute in San Diego, Woolgar and Latour (1979) turned their ethnographic lens from the myths surrounding so-called exotic tribes, to what they referred to as the tribes of scientists. A key finding for them was that scientific facts do not emerge fully formed from data or the minds of singular researchers; they are the result of so many things from scientific training, to political agendas, and current dominant scientific paradigms – not to mention funding streams. This line of thinking is not anti-science, as many of his detractors suggest, but rather it is an effort to demystify scientific thinking, to open it up and have a look inside, and recognize that it is possible anywhere (cf. Mavhunga 2014).

If we reflect on Heart of Darkness, Latour's work offers a potential explanation for Marlow's descriptive lapses – if we take the generous view that Marlow is more ignorant than he is a racist. Marlow is unable to describe the human ecosystem and its attendant social relations around him because he lacks cultural framing to understand and even describe his indigenous counterparts. Had he adopted the basic Western ideology that the Mbuti, the Baka, and the Batwa were subhuman he may not have considered their place warranted in the text. Our reason for raising this point is that the perceived universality of Western Science is a myth. This is a myth that remains to this day. Nowhere is this clearer than in the development policies of perpetuated by the United States and its allies, China, the Japanese, Russia, and other historically dominant powers.

In the discussion above we argued that the 'Orient' is no more exotic than the 'Occident', especially in terms of forms and practices of inquiry. Indeed, when we turn the ethnographic lens back on ourselves (i.e., western scientists, engineers, and social theorist) it becomes clear that our own practices are as chaotic as those which we call primitive, traditional, and indigenous. Let's explore this through a vignette brought from East Pakistan and the International Bank of Reconstruction and Recovery, which is now known as the World Bank.

East Pakistan: The Karnaphuli Pulp and Paper Mill

In 1950, after partition and independence, the newly formed Pakistani government approached the International Bank for Reconstruction and Development about funding for a pulp and paper mill to be sited along the Karnaphuli River in East Pakistan (now Bangladesh). The project was innovative in that it would use local bamboo as its raw material.

The mill was designed to produce 30,000 tons of paper per year with capacity for expansion depending on the needs of the market. The raw material came from a bamboo reserve upstream from the plant. The company secured a long-term lease with the East Bengali government. Early tests at the mill determined that the paper produced there was the same quality as paper produced from white pine in the United States.

The paper was primarily produced for export since the need for the type of paper produced at the mill was limited in East Pakistan. To support plant operations a large labor force would have to be brought in. The company created infrastructure, built a hospital, a school, and over 1,000 living spaces to support the imported labor force. The processing facility was completed in 1953 and the housing and other infrastructure in 1954–1955. Economists at the International Bank expected the return on investment to be between 18% and 24%, based on assumptions of 80–100% of total plant capacity.

Despite the promising production numbers, the plant suffered its share of managerial and technical problems. For example, the mill suffered labor problems because the mill workers were recruited from local Bengali workers while non-Bengali Muslims from India and West Pakistan made up the management team. In 1955, tensions erupted between the two factions leading to riots and the assassination of the general manager of the mill, who had been the project's architect and key promoter.

The plant also ran into technical problems. Equipment for the plant was brought in from Switzerland, Italy, and the United States. Because of the different chemical content of the bamboo the chipping, pulping, and chemical recovery processes were different than expected. After all, these machines were designed for applications in the pine forests of Europe and North America and not bamboo forests of South Asia.

Finally, after a few years of stable operation the mill faced another obstacle – 85% of its raw material, the bamboo in the reserves, flowered and died, making them economically worthless (Hirschman 1964). While science knew bamboo flowered every 50–70 years no one considered if the dead material could be used in the production of pulp. Unfortunately, it could not; it disintegrated during the downriver transportation process.

The case of the Karnapuli Paper mill illustrates the points we raised earlier. The first point could be described as Muslims are Muslims, period. Without considering the colonial history of Great Britain, the project's architects assumed that having workers who shared a faith could work together harmoniously. There was no effort to account for a common practice across the former British Empire, which was to import bureaucrats and administrators from abroad without regard to ethnic tensions. Or, to turn different ethnic groups against one another to promote in-fighting rather than outright revolution. For example, during their occupation of what are now India and Pakistan, the British employed a tactic used by many colonizers – think the French in Rwanda when they made people identify as Tutsi, Hutu, or Twa – by establishing religious affiliation as a set of silos to divide what was a tradition of diverse thought. Aggravating this further was the 'one unit' scheme enacted by the government based in West Pakistan, which was the home of political, spiritual, and economic power.

Another challenge to the project was the failure of scientists to realize where the bamboo was in its lifecycle. A massive effort was marshalled to determine the potential productivity of the region, whether the bamboo product was priced competitively, and finally, would the product appeal North American consumers, where the primary market was to be. Yet, none of these economic factors really shaped the outcome of Karnapuli. Rather, it was the bamboo's biological clock. Like apples and the law of gravity, the biological clock of a bamboo forest cannot be reset, turned back, or brought forward. One could argue that it was the "ideology of [post] imperialism" that shaped bamboo science.

Here we have shown that western science can be chaotic and guilty of asking the wrong questions. Furthermore, the so-called primitive worlds had complexity and nuance that science was incapable of apprehending. In addition, and this is what we examine in the next section, scientists and engineers failed to understand the cultural entanglements, between and within different social groups, including the colonizers themselves. In the next section we examine the entanglement of Kwame Nkrumah's Pan African vision and his development approach in the early years of the Republic of Ghana.

Homegrown Development in Ghana?

As an incipient nation state, Ghana realized massive profits from the global cocoa boom. On this foundation, the government of the nascent country sought to position itself in the global economy and as a model for the rest of the Continent to follow. In terms of infrastructure, President Kwame Nkrumah had two signature projects: the Akosombo Dam that sequestered 148 km^3 of water and was, at the time, the world's third largest hydroelectric complex. The dam provided electricity for Benin, Togo, and Ghana. For Nkrumah, it was to be a catalyst for supporting Ghana's industrial aspirations. In particular, it would support the energy needs for processing bauxite into aluminum for Ghanaian US business partnerships. This industrial center was central to Nkrumah's other signature project the deep-sea Port at Tema. In Nkrumah's vision of an industrialized Ghana, the new port was to be a hub for industry and the point for importing and exporting materials and value-added products.

Developing a port at Tema was first envisioned by the British colonial government. It was not until independence that the port project was realized and then commissioned by the Ghanaian government, in 1962. For our purposes, there two parts of the project are important. First, unlike the Akosombo Dam (also known as the Volta River Project), whose design and construction were put out to tender by the Ghanaian government. The Akosombo Dam was designed by a British architect and developed by an Italian firm. In contrast, the Port at Tema Project was decisively 'Ghanaian'. It was designed, planned, and developed by Ghanaians. In particular, it was Theodore Clark, Ghana's first trained architect who envisioned the port, industrial, and residential areas that were to be part of the port project.[1] The port was to be the

1 Not surprisingly, it was similar to the 'Garden City' first inspired by British architect Ebenezer Howard.

basis for Ghana's effort to break from its colonial past and create economic opportunities based on self-sufficiency and global connectedness.

Despite the fact that Tema was a 'homegrown' and built in the spirit advocated by Nkrumah to create a 'new culture of shared belonging' that, in his mind, was required for independence, it aggressively challenged Nkrumah's efforts at modernization, as well as revealed that rather than an icon of pan-African liberation, it represented a kind of Trojan Horse. Unlike the Trojan Horse at Troy, the one at Tema did not directly bring troops and violence to the project; rather when opened it revealed the entanglements and dissonances of Nkrumah's goals and very western ideology he sought to escape. Ghana might have been the first independent state to experience this,[2] but since then, the power of these ideologies has plagued former 'Third World' countries around the globe, in general, and sub-Saharan Africa countries, in particular. Let's explore the Tema case a little further.

The village of Tema, before the port, impounded 3.9 million m^2 of land. Tema became referred to as a 'long-forgotten Ga fishing village'. The Ga are an ethnic minority in Ghana, Togo, and Benin, who settled along the Gulf of Guinea in the fifteenth century after migrating from a region that is part of what is now called Chad. For the Ga, it was not a 'long-forgotten fishing village' but one of their original settlements and an important spiritual place that dates back half a millennium.

Like their neighbors, the Akan, the government does not have well-defined property rights to Ga land. Complicating this further, is that the Ga do not specify property rights to individuals, but they circulate through local power structures that include chiefs and local priests. The priests do not preside over the land in service of the Ga people but act as custodians of the land for the gods. Furthermore, the gods inhabit these lands as foretold by certain geographical features. In Ga religion it is forbidden to remove or alienate the gods from their land. There is no ritual to do this like in other regional religions. Thus, the communal gods have ultimate ownership of the land – it is not for people to give away or sell. The Tema gods were especially important because the land that Tema occupied was between two lagoons, and to the Ga, this represented the cycle of seasons and the human relationships to nature. For the Ga, disturbing the land in the way that the government laid out could jeopardize the group's well-being.

We know how this story turned out. Tema became the industrial estate that never was. The plans for manufacturing and adding value to local resources never materialized, though the Port of Tema was realized and became an important trade zone for Ghana. Nkrumah's approach further illustrates the social construction of, and instrumental rationality, attendant to science and engineering and development. The shared culture of belonging advocated by Nkrumah came at a cost to the Ga. More

2 Ethiopia, having never been colonized, was the first sub-Saharan African country to experience this through the modernization efforts of Haile Sellassie, the country's former Emperor.

importantly, it showed how values can be reconfigured to meet the perceived needs of the day. Finally, for Nkrumah, the appeal of the industrial district, though strong-arming Ga culture into submission, was that it would bring economic prosperity and modernization to Africa's shining Black Star. Again, culturally exogenous and indigenous cultural tensions stemming from the 'western way'.

With a different set of goals in mind, Mavhunga (2017, 2018) examines the social construction of science, technology, and innovation in his home country Zimbabwe. Here Mavhunga sees science as a social process. This is a fact he celebrates rather than shies away from. His accounts seek to provide African forms of scientific (STEM thinking, if you will) with the legitimacy they deserve by elucidating the Shona language of science and familial and ancestral practice of orally sharing scientific traditions. Again, as observers we cannot see what's happening before our eyes. Our ways of doing, our definitions of what counts as knowledge, have, through the teaching and practice of our ancestors have blinded us to other cultures of inquiry.

Overview of the Volume

In the pages that follow, the contributors of this book seek to provide solutions to these issues conceptually, methodologically, and in practice. It supports the points we have raised here that development is more than complicated; it is complicated by histories, actors, traditions, technology, cultures of inquiry, and invisible hybrids and entanglements. Rather than exploring these concepts in more depth, we explore thoughtful solutions to these issues in a language that is comprehensible to those outside the social sciences or humanities. The chapters represent a transdisciplinary vision and bring it straight to practice, in terms of technological ideation and design, policy ideation, and evidence-based policy approaches, to educational perspectives and leadership skills required to navigate the complex network of science and technology as it is applied to global development, or any other context, really.

The book has four parts. The first will be familiar territory if the reader has experience field of development. If not, don't worry, we have intentionally presented this work in a rich, but accessible way so that the uninitiated can follow along, while the experienced reader will understand the depth and broader value of the work. In the first part, which comprises the first two chapters, we seek to frame the history of applied science and engineering practice as it was conditioned by beliefs at their time of deployment. This includes World Bank projects but also what some have called homegrown alternatives. Regardless, what we see is an entangled mass of ideas, ideologies, assumptions, and values that have been mobilized from around the world and brought to bear on plans and projects. For example, Kwame Nkrumah, the father of Pan Africanism, doesn't create a positively *African* subject in an independent Ghana. Rather, he replicates the hubris of both *homo-economicus* from the west and the obsequiousness of *homo sovieticus* as it he understood it from the Soviet Union. We wonder if perhaps the 'paradigm' isn't the way to scaffold this analysis? Feminists, Afrofuturists, postmodernists, and others have implicitly, some explicitly, suggested that dichotomies are themselves a fabrication of dominant worldviews. Accordingly, our primary interest when seeking socio-technical change should not be trying to define the old in reference to the new or vice versa. This is the classic red herring. We explore this further through an examination of so-called do-gooder experts (Chapter 2). We take the discussion to the actors in science and engineering for development and describe how they and the beneficiaries of their brilliance are both actors with agency in shaping interpretations and implementations.

The second part of the book exposes this multidimensional agency, and instead of trying to cover it up or just ignore it, we embrace it through transdisciplinary thinking and design for justice. We show that traditional approaches to science and technology for development have reflected the 'spectator theory of knowledge' in what has often resulted in the blood sport of development. For example, engineers 'see' a problem: the lack of economic opportunity, the need for a bridge, or electrification. Yet, 'problems', and especially development ones, have pasts and futures. As we point out in later chapters, for example, economic development in East Pakistan and Uruguay created the material accoutrements for development, but did not envision why these problems were perceived in the first place or the unintended consequences of 'fixing' them on the groups they sought to help because of the hyperfocus on economic development and efficiency. Next, we introduce the notion of design as inquiry. Doing so helps recognize a tension among 'what is', 'what could be', and 'what ought to be'. As the latter complicates straight lines drawn between 'what is' and 'what could be', design becomes an inquiry without a predetermined answer. It invites the engineer to be an ethical participant in the creation of technologies that serve the common good. These are elements of a new paradigm, for sure. Again, our goal, as it turns out, is not to create new intellectual giants for the taking, but to shine a light on the complex dimensions of doing science, design, and innovation, in development.

The third part brings these concepts to STEM education for development. By exploring leadership, ethics, the role of 'nonexperts', and training students to lead with empathic inquiry all offer ways forward as we educate future generations. Given the global retrenchment on empathy and celebration of difference this would seem to have many different pedagogical and educational opportunities. This part is for the student, educator, and practitioner alike in that it provides best practices for teaching the essential skills (what some call 'soft' skills) for a new techno-social leadership in development.

The fourth and final part examines different 'scientific' dimensions for framing the use of technology in development and policy interventions. Chapters 8–11 decenter conventional thinking and explore technology for migration, policy evaluation, poverty assessment, and environmental robotics. Chapter 8 points out the continued challenge of policies that seek to sustain the dichotomy of 'right' and 'wrong'. In doing so, policy designer provides information on systemic issues but fails to tell the true story about the challenges, choices, and desires of the people whom the policies impact. To address this, the authors build from Chapter 3, where policy design is an invitation for inquiry. The authors advocate for mixed method approaches that enable the experiences of the policy beneficiaries to emerge, not the preordained assumptions of policymakers. Chapter 9 examines the most conventional, most relied upon approach in policy and program assessment, the randomized controlled trial (RCT). The authors deconstruct the RCT and demonstrate why these forms of *pure science* must also extend transdisciplinary design approaches. Further, they argue that RCTs must move beyond the domain of the 'expert' and promote integrative thinking and egalitarian processes. Chapter 10 looks at one of the newest tools being brought to the science and engineering for development toolbox and artificial intelligence tools. Using the context of poverty assessment through data science, in general, and machine learning, in particular, the authors reveal that our shiniest new tools are bound to replicate the failures of the past unless we understand poverty from the people who experience rather than what global institutions think it to be. We must train our machine learning models, for example, on data that was produced to assess poverty, and most importantly, poverty through the 'eyes of the poor'. Chapter 11 examines robots and their tools for addressing development. In general, the authors are skeptical of robotics in general. But, when they focus on a subfield, environmental robotics, they reclaim the egalitarian design approach called for in earlier parts of the book. These chapters provide an empirical basis for the concepts developed in the previous chapters. What are examples of socio-technical entanglements and what value can we bring to opening up paradigms for shaping by diverse voices, from a range of background and experience, in peer-to-peer negotiation.

Laureen Elgert

Chapter 2
Do-Gooders, Tyrants, and Anti-politics: Experts in Development from Postwar to Post-development

Introduction

The introduction of this book sought to disentangle and decenter western notions of science and engineering. Another way of saying this is that the West has experts where those elsewhere in the world are passive players in our world of rarified scientific and technological innovation. In few areas of study and practice, have 'experts' and their expertise received as much attention as in development studies. This chapter will examine three positions on experts from three paradigms of development, each responding to political imperatives and failures of what came before. The purpose of this is to promote a more critical analysis of both unquestioningly and absolute acceptance and rejection of expertise in development policy and practice, as elsewhere, and create an increasing openness and self-reflection among development engineers.

In the postwar context, development experts were mainly concerned with edging newly independent countries toward growth and modernization through the 'neutral' application of engineering and economic principles for large-scale infrastructure development and economic restructuring. In the 1990s, participatory development experts facilitated partnerships with local people, considering their unique expertise in terms of their own experience and aspiration. Finally, some critics of historical development began to advocate for a new era of post-development and did away entirely with the idea of external experts and the notion of development altogether, understanding it as nothing more than a new phase in global colonialism, and a project of power and control. The chapter concludes with some thoughts on the current and evolving perspectives on experts in the age of late-stage capitalism, entrenched globalization, and ubiquitous environmental crisis.

Postwar Development: Political Origins and Technocratic Intervention

International development emerged as a distinct area of scholarship and practice in direct response to the need for new modes of thinking around how to facilitate eco-

https://doi.org/10.1515/9783112218716-002

nomic growth and industrialization in a postwar, postcolonial global era. The postwar Marshall Plan for European reconstruction was signed by the US congress in 1948 and together with the creation of the Bretton Woods institutions, including the International Bank for Reconstruction and Development and the International Monetary Fund, is widely taken as the dawn of global development. The institutions were designed to finance and facilitate the rebuilding of the physical and economic infrastructure of war-ravaged Europe, to promote international cooperation, and to ensure that the continent fell on the 'right' side of cold war politics. In this postwar context, dozens of countries in Africa and Asia achieved independence and provided another impetus for Western intervention. Both the tug of cold war between the US and USSR, and the economic troubles in Europe and the rest of the world, created and exacerbated by urbanization, health and governance crises, and population growth, provided incentives for international assistance.

The Western imperative for political and cultural influence drove the emergence of a global development enterprise. Despite the distinctly moral and political undertones of development that were overt and explicit in early 'development discourse', post-cold war attitudes toward global development interventions and social progress sat comfortably alongside technological hubris. While technology was front and center in debates and discourses of political superiority, technology itself was viewed as a self-evident marker of advancement and a morally righteous means of achieving social progress. The coupling of technological and social progress is epitomized by the now famous 'kitchen debate' when the US President Richard Nixon and the USSR General Secretary Nikita Khruschev, standing in a model of a modern tech-laden kitchen (which Nixon held as a symbol of the good life), sparred about the merits of communism and capitalism. While the episode is often recounted with whimsy, it illustrates the way in which technical expertise at the time (and arguably still) was revered and equated with development and progress, but was also couched in the political struggles of the cold war.

Infrastructure for Takeoff

At midcentury, development thinking was largely based on social and economic theories that contrasted non-modern with modern societies and posited that 'progress' toward economic growth and industrialization was a linear process whereby societies would pass through a series of stages in predictable, systematic, and universal ways. Modernization was seen in interrelated sociocultural terms (social organization, traditions, family structure, and cultural institutions) and economic terms (the means and order of production and consumption). This thinking was propagated by sociologists and economists of the day. Talcott Parsons, one of the most influential sociologists in the United States, argued that naturalistic concepts could be used to analyze society, such as organic evolution, variation, selection, adaptation, differentiation, and integration, could be applied to questions of societal change from superstition and culture to rationality and reason (Parsons, 1964). In other words, social characteristics and values were hierarchical (modern is superior and more developed), universal (all societies progressing to a common end), and inherent (the differences are natural and objective). Of course, the contested, yet oft assumed synonymity of modernization and westernization in the mid-century (Lal, 2000), makes it clear where legitimate expertise 'lives'.

It does not take a stretch of the imagination to see the influence of this mid-century sociology in the work of Walt Rostow, the economist and father of 'modernization theory'. In 'The Stages of Economic Growth' (Rostow, 1959), he mapped a linear, phasic pathway toward development that he used to explain varying degrees of economic development and underdevelopment in nations across the globe, and more importantly, how economic growth and development could occur. But there was also a prescriptive element to the stages. With the right kinds of investments and expertise including the encouragement of western cultural values and industrial policies, argued Rostow, nations could move from being traditional societies to a transitional phase toward modernity and industrialization.

Infrastructure development during the 'transition phase' was one prerequisite that would enable countries to 'takeoff' toward modernization (Rostow, 1959). This mantel was in part taken up by the experts at the World Bank, who assumed the role of 'Builder and Engineer' to the world. The wisdom of these mid-century experts went more or less unquestioned and was focused on controlling nature and building infrastructure. Indeed, nation building and the symbols of a powerful and progressive nation often took the form of large-scale, awe-inspiring, 'charismatic' infrastructure development. Dams, for example, became a signature World Bank investment, starting with the first World Bank loan to a non-European country to Chile in 1948 for a hydroelectric dam. The World Bank became the largest funder for large dam construction in the world, providing billions of US dollars for hundreds of dams, in nearly 100 countries. More broadly, dams became a staple of national development plans (Altinbilek, 2002; Khagram, 2018), as did other massive infrastructure projects such as new highways, with the intention of bringing progress toward modernity, and economic growth to less developed nations.

By the late 1970s and early 1980s however, critical observers had much to say about these projects that highlighted exaggerated benefits, underestimated price tags, and the disproportionate costs borne by the (often rural) poor, including displacement of rural communities, cultural erasure (Kala, 2001), and environmental degradation. Such critiques underlined the unintended, yet predictable consequences of the overreliance of technical expertise in development without adequate attention to environmental context and to diverse stakeholders. Two large infrastructure projects became emblematic of the rise of this kind of dressing down and erosion of the hubris of

World Bank experts as builder: the Narmada Dam project in India; and the Polonoroeste highway pavement project in Brazil.

The Sardar Sarovar dam, the largest in a series of 3,200 dams planned along 1,200 km of the Narmada River, known as the Narmada Dam project, broke ground in 1961, but construction gained momentum with World Bank funding and involvement in 1979. In addition to creating hundreds of jobs, the dam was expected to generate cheap, renewable energy for urban and rural areas alike, and provide water for irrigation projects in drought prone areas of Gujarat. But massive protests erupted over the dam and were brought to international attention by celebrities like author Arundhati Roy, and local and international NGOs. The opposition charged that more than 2,00,000 people would be displaced from mainly agricultural villages up and down the Narmada from both flooding and losing water supply altogether. The Environmental Defense Fund called the dam project, 'India's greatest planned environmental disaster' http://websites.umich.edu/~snre492/Jones/narmada.html

Another challenge to the primacy of the World Bank's development experts set environmental activists, Indigenous peoples' advocates, and even the weekly US news program '60 min' against the Programa de Desarrollo Integrado del Noroeste (POLONOROESTE) Project in Brazil. The World Bank joined as the only non-Brazilian investor on the project in 1981, aiming to pave 1,500 km of dirt roadway connecting Brazil's southern commercial center of Sao Paolo and the state of Acre, in the furthest western reaches of the Amazonian region. The paved road was intended to expedite economic growth and development in the Amazonian region by expediting the transport of farmers and settlers into the region, and resources and commodities out of the region. The massive scale of migration that followed was unanticipated, however, and occurred before authorities were able to demarcate and set up protections for indigenous lands and institutions to promote 'sustainable agriculture'. The international community buzzed with criticism as the threats to the physical and cultural survival of indigenous people, and the fires clearing forests for agriculture hit international airwaves.

In response to national and international outrage, funding was pulled from both the Polonoroeste (in 1985) and the Narmada Dam (in 1993) projects, and scholars point to this acquiescence as key moments of reckoning in the development of critical discourse surrounding the role, authority, and legitimacy of (technical) expertise in the World Bank, and more broadly in development discourse. In both cases the World Bank hadn't considered or given enough credence to local conditions and potential, but predictable, unintended consequences of technical expertise in context: "Suddenly the Bank found itself defined as doer of harm and teller of lies, and required to react to outsiders' ideas about how it should do its business. For an organization that had always prided itself on its service to humanity and unrivalled technical expertise, this was bewildering" (Wade, 2016, p. 215).

While the focus on large infrastructure did not come to a halt, and continues to this day (some say large infrastructure is even enjoying a resurgence at the World

Bank (Bosshard, 2013), a combination of stinging critiques and shifting priorities turned attention away from investing in national monuments toward restructuring national economies, as the World Bank went from 'builder' to 'banker' to the world.

Economists and Universal Economics

The Latin American debt crisis of the 1980s set the stage for perhaps the most profound example of development expertise as tyranny the postwar world has known. As American banks amassed huge deposits of oil money from the Middle East, they ran out of places to lend. While corporate loans did not provide sufficient borrowing capacity, sovereign nations, many in Latin America, addressed this dearth of borrowers. Between 1975 and 1982 Latin American debt to commercial banks increased cumulatively at 20% per year, and interest payments grew four times, to around 50% of the region's GDP. At the end of 1970, total outstanding debt from all sources totaled only $29 billion, but by the end of 1978, that number had skyrocketed to $159 billion. By 1982, the debt level reached $327 billion (Federal Deposit Insurance Corporation, 1997, found in; Sims, Federal Reserve Bank of Chicago, Romero, and Federal Reserve Bank of Richmond, 2013). Much of this debt was taken on by dictators or unaccountable leaders for personal or nefarious purposes.

In August 1982, Mexico was no longer able to service its loan payments and defaulted, sparking concerns that other countries would follow. The IMF declared an international debt crisis. Despite that much of Latin American debt was accumulated due to a need to lend more than borrow, and that much of the loan revenue was siphoned off into Swiss bank accounts; some built luxury hotels, some of the money was used to fund militaries to oppress their own people. Yet citizens were held responsible for repaying these 'odious debts', as they came to be known, and rationalized some of the most harmful expertise of the twentieth century: structural adjustment policies (SAPs).

The crisis opened an opportunity for economists, as the World Bank's 'primary advisers' (https://www.worldbank.org/en/archive/history#), to take the driver's seat in global development. A small group of policymakers, economists, and academics gathered to debate policy responses to the economic crisis in Latin America, and it was here that think tank economist John Williamson's 10-point contribution to the debate called 'The Washington Consensus' got attention. He proposed helping countries qualify for new loans and debt refinancing by restructuring their economies, through a standardized set of conditionalities that were broadly described as 'embracing free market principles': market liberalization, privatization, deregulation, and disinvestment. Specific policies mandated measures such as deregulating industry (removing labor and environmental regulations, for example), opening borders to foreign investment, selling off public assets such as utilities and public lands/natural resources to private interests, and imposing austerity measures (reducing or eliminating subsidies and investments) in areas of health, education, and agriculture. Free market principles also dictated a preference for exports such as cash crops, often at the expense of food production for domestic consumption. Indeed, experts assured that this 'tough medicine' would accelerate economic development, stimulate innovation and ingenuity, and generate new economic growth that would trickle down over time and benefit the masses. These SAPs and the framework of neoliberalism that they operationalized, were the single most important and defining policy framework of development thinking in the 1980s (Gavin, 1995), and remained so over the next four decades.

Neoliberalism ushered in an unprecedented rule of experts in public life and the private lives of citizens; expertise became profoundly ubiquitous, legitimized, and depoliticized. The neoliberal rationale drew a sharp distinction between knowledge and politics; experts (largely economists) who lay claim to objective knowledge about the economic life of nations, could claim to operate outside the realm of values and politics (Collier, 2017). What was remarkable about the Washington Consensus, and neoliberalism more generally, was the universality with which economists believed in the universality of their prescriptions and practice. In fact,

> The existence of development economics, based on different assumptions and methodologies than the ones used in the rest of economics, was denied on the grounds that economics is a universal 'science', with laws that apply across time and place. In this view, development economics is distinguished from other (more legitimate) branches of economics only by the fact that the data sources used happen to come from developing countries. (H. J. Chang, 2003, p. 4)

But the Washington Consensus never earned consensus. Structural adjustment was immediately condemned by the public and analysts alike, and over the past few decades has inspired a staggering body of critique, much of which points to outcomes that are decidedly, un-evelopmental. SAPs concentrate wealth and exacerbate inequality (Forster, Kentikelenis, Reinsberg, Stubbs, & King, 2019). They worsen health outcomes and limit access to education . . . SAPs have been linked to increasing deviant and crime rates, ethnic hatred and discrimination and welfare problems (Rono, 2002). Furthermore, SAPs failed to improve national economic performance indicators, their explicit purpose, and contradicted the experience and evidence from rich countries: that protectionism, subsidies, and industrial policy (all things that go against the SAPs) were required for their economies to grow. Indeed, the SAPs cultivated poverty and stagnation, by 'kicking away the ladder' to industrialization, growth, and prosperity (H.-J. Chang, 2002).

Stinging critiques of the wisdom, or lack thereof, of universalized expertise – engineering, economic, and otherwise – drew attention to power relationships between development experts and 'subjects'. James Scott's classic 'Seeing like a State' (Scott, 2008), for example, takes direct aim at grand, hubristic endeavors – even those (perhaps especially those) that are driven by intentions that seem benign. Scott argues that state projects of modernization, while couched in terms of progress, often have disastrous outcomes that are widely condemned, but also more insidious outcomes. Development expertise that is harnessed by cutting through landscapes with highways, designing the 'rational' city, and resettling villages, is not primarily exercised in order to improve society, but to make society and its citizens legible to the state, further embedding state power and control.

The Participatory Turn: New Experts in New Places

The unintended consequences of large infrastructure projects, and the intended but partial consequences of SAPs, created tyrants of the development establishment. Experts as builders and experts as bankers largely failed in producing any substantive improvements in global poverty, and in the 1990s this led to a reconsideration of expertise more generally, as "The ineffectiveness of externally imposed and expert-oriented forms of research and planning became increasingly evident" (Cooke & Kothari, 2001:5). From the ashes of what came to be known as the 'lost decade' in development, the participatory turn in development studies (Robert Chambers, 1986) aimed to mitigate some of the damage done by neoliberalism's experts. Participatory approaches and methods promised to address the critiques of top-down, expert driven development, by putting local knowledge and community perspectives at the center of development research and practice. The 'participatory rural appraisal' or PRA as it came to be known, was pioneered by Robert Chambers as a response to these critiques, and pushed development practitioners to involve local people, their knowledge, priorities, and preferences, at the forefront of development interventions (Robert Chambers, 1994). Today, this sounds laughably obvious. But when Chambers' Rural Development: Putting the Last First (1983) was published in 1983, PRA, and participatory approaches more generally, flipped the notion of expertise on its head and provided (at least theoretically) a radical departure from convention development.

Ultimately, Chambers argues for a 'new kind of professional' who is expert at processes and engagement; who understands the value of local knowledge and has skills in methods to tap into it. This new development expert is a partner in, rather than the architect of development. Experts could be abided, but in a mainly consultative role, and were more appreciated as facilitators, with expertise in listening, bridging, and empowering processes, than as project managers (Craig & Mayo, 1995). Chambers' own positionality stands in contrast to the benign hubris of the development expert of the past. He identifies himself as a learner, not a teacher, and is self-effacing, introducing himself in the preface of 'Putting the Last First' by saying: "And let no one suppose that my position is 'holier than thou'. The book is addressed, in some embarrassment, to myself as much as to others" (Robert. Chambers, 1983: preface).

The participatory turn was headlined by internal critique of development and conventional development experts that focused on the inefficiencies and ineffectiveness of top-down interventions, as outsiders misread local landscapes, are blinded by biases, overlook important realities, and disregard community dynamics that might render even the best ideas useless. Participatory approaches pushed development experts to examine their own biases in order to overcome them and more effectively allow programs and initiatives to be guided by local wisdom. Seasonal biases, for example, may arise because experts rarely visit during the rainy season, when travel is more difficult, or a harvest time (Fortmann, 1985). Gendered biases arise when women's work is overlooked, and men take a lead role in interacting with outsiders (Elson, 1995). Cultural biases often inhibit an understanding the role of local institutions in environmental conservation or maintaining social cohesion (Grillo & Stirrat, 1997, 2020). Urban bias may hamper rural development through disproportionate influence and political control over decision making (Lipton, 1977, 1984). Such biases often stem from an inability or an unwillingness of the development expert to become immersed in local reality to understand the local context and the priorities of communities. A more appropriate form of expertise from the perspective of participatory development, is knowing how to channel local knowledge and challenge or overcome bias, to achieve a more 'local' view of reality.

Meanwhile, those at the World Bank were acutely aware of critiques of its failings, and were listening to the debates about bias and participation. In a partial attempt to repair the Bank's reputation, so damaged by the abject failure of SAPs, they responded with *Voices of the Poor*. *Voices of the Poor* was a series of efforts through the 1990s that focused on collecting and analyzing vast amounts of qualitative data on the experiences of the world's poor, collected through interviews with people in 47 countries. Culminating in the 2000 report, *Can Anyone Hear Us?* (Narayan, Chambers, Shah, & Petesch, 2000), the interviews repositioned the poor as the 'real' poverty experts, and sought their perspectives to (re)define development and the 'good life', to understand peoples priorities, and to reassess the nature of relationships and interactions between the poor vis-à-vis the market, the state, and civil society. This work was subject to extensive critiques of tokenism and lack of 'rigor', but guided by a philosophy and led to findings that represented (and in part, perhaps, led) nothing short of a sea-change in the way that the development community defined poverty, development, and the range of actors who were considered experts in the field (Narayan et al., 2000).

From its roots in rural development, participatory approaches took flight in many areas of development, following the contours of the 'governance turn' in 1990s development studies more generally – the idea that good (more democratic) governance had been the key missing ingredient in the development agenda leading up to the 1990s. Fostering democratic government institutions was of interest, but this idea also filtered into more niche areas of development practice. Participatory GIS, for example, aimed to address issues of "uneven access to GIS and digital spatial data and diversify the forms of spatial knowledge and spatial logic that may be incorporated in a GIS" (Elwood, 2006: 693). Early work around participatory budgeting emerged from

Brazilian cities such as Porto Alegre, Belo Horizonte, Sao Paolo, and Belem, where citizens were invited to public forums where they would deliberate on local priorities, and collectively make decisions on resource allocation in the public budgets. "These programs confront Brazilian political legacies of clientelism, social exclusion, and corruption by making the budgetary process transparent, open, and public" (Wampler, 2000, p. 2). A substantial body of research was conducted around these so-called participatory experiments and the outcomes seemed mixed, and dependent on a variety of contextual characteristics (Wood & Murray, 2007). All of these projects, however, illustrate the location of expertise (on the most pressing and effective uses of budgetary resources) in the citizenry, and external consultants as facilitators.

NGOs: Locus of Grassroots Expertise and Advocacy?

In early development thinking, expertise was largely concentrated, among a small group of centralized professionals (largely economists) and government authorities for whom they consulted. But the NGO-boom, or the 'NGO-ization of development' of the 1980s expanded and decentralized ideas of expertise (Alvarez, 2009; S. Roy, 2015, among others). Despite their variety, the very name of nongovernmental organizations emphasized that they were defined in contrast with top-down and sometimes authoritarian governments, as producing, deploying, and embodying 'local knowledge', and providing a democratizing force in society (Bebbington, Farrington, Lewis, & Wellard, 1993). The boom itself refers to the expansion of the NGO sector from a handful in the late 1970s, to tens and hundreds of thousands by the late 1980s. The expansion was enabled, perhaps necessitated, by the roll back of the state and other structural adjustments, and development's emergent focus on governance and democratization. NGOs were understood to represent civil society (which became the term to refer to the NGO sector) and deploy specialized knowledge in both spatial (local) and sectoral (health, education, livelihoods, etc.) terms.

Despite the common view of NGOs as benign, local, and representative, scholars point out that they are a fundamentally privatized form of social and political action, and are often funded internationally to promote agendas that are not necessarily locally determined or sanctioned. Indeed, dozens of forms of NGOs exist today, from large, heftily funded, and influential global NGOs (INGOs), to grassroots, community-based organizations (GNGOs). The professionalization and mainstreaming of NGOs (PNGOs), particularly in the context of the shrinking state, has been criticized for the deradicalization of resistance (A. Roy, 2014) and as extending the ethos of neoliberalism deeper into civil society (Choudry & Kapoor, 2013).

The Return to Tyranny

Following Chambers, a wave of research reported on the importance of 'participation' – though what constitutes authentic and legitimate participation, as opposed to mere consultation, or worse, manipulation, remains deeply contested. In 2001, Bill Cooke and Uma Kothari published the much anticipated and well-received set of essays called Participation: The New Tyranny? Which consolidated the calls for called for renewed critique of development expertise, this time in its participatory form. The strong wording in the title was intentional, and meant to call attention to the failures of 'participation' to empower citizens and activate appropriate development. The tyranny here referred to that of 'decision-making and control' whereby participatory facilitators override existing legitimate decision-making processes; of 'the group', through which group dynamics lead to participatory decisions that reinforce the interests of the already powerful; and of 'participatory method' that had largely driven out other methods which have advantages that participation cannot provide. There were echoes of these arguments in later interrogations of participation that found that often participatory processes were captured and dominated by wealthier, more educated, higher status, politically connected men (Mansuri & Rao, 2013).

Post-development: Alternative Development or Alternatives to Development?

Once seen as the solution to the tyranny of universal expertise, participation had itself become the tyranny, and created a growing discontent with the development 'industry', or 'complex', as it had become known, writ large. At the center of the post-development paradigm is a wholesale rejection of conventional external experts (Bennett, 2012), and development more generally as ineffective (at best), or more critically, as an embodiment of power and influence. Debates around post-development continue, and range from reformist to radical. Scholars argue between alternative development, on the one hand, and alternatives to development, on the other. Discussions around alternative development revolved around 'new ideas' (and new expertise?) such as microcredit and sustainable development, or more generally about institution building and good governance. But somehow these ideas never moved too far from the spirit of the 'development panacea' – the idea that 'underdevelopment' was the lack of a certain something, and once that could be discovered, the pandora's box of growth and progress would open. And while the hubris of external expertise was toned down from the dominance of World Bank builders, it was still detectable, and development was essentially something that came from the outside. In his review of Easterly's *The Tyranny of Experts* for example, French outlines the persistent attachment to experts and expertise, independent of type or content:

> The emphasis on technical solutions, from tariff regimes to industrial policy, has empowered experts, who have gradually come to constitute a formidable self-reinforcing constituency. Even when they disagree on the details, as often happens, they seem unanimous in their belief that what poor countries most desperately need is outside advice. (French, 2014)

Easterly – a 'recovering' career expert at the World Bank – argues that this focus on technical solutions, often originating in the West, fails to address the root causes of poverty, and has frequently led to an inadvertent buttress for autocratic leaders who trample on the rights of the poor, stealing away attention from what is really important. Individual rights and freedoms, not external expertise, he argues, is the more promising model for development (Easterly, 2014).

Alternatives to development involve more radical appraisals of twentieth- and twenty-first-century development experts as the purveyors of anti-politics, depoliticizing what at its heart is fundamentally a political project of power and control. Such approaches critique not just how development has been carried out, but the fundamental concept: its historical roots and entanglements with colonialism, westernism, and capitalism. While much of the work done under the participatory development framework, aimed to empower the poor, advocates of post-development seek emancipation from development altogether. Here, the universalism of development expertise is a tool for suppressing independence (Illich, 2003) and oppressing the subaltern (Saffari, 2016). Development is better understood as neocolonial than emancipatory, and the expert as colonizer, albeit under the well-worn, but debunked guise of a neutral technocracy (Escobar, 2011). These positions are born out by research that charts how development is increasingly linked to the security concerns of the West.

While even moderate, mainstream development consultants now recognize the value of 'local knowledge', more radical post-development critiques (granted, post-development contains multitudes, see Corbridge, 1998) turn toward a wholesale redefinition of expertise as inherently local and all nature as situated within a historical, cultural, and political time and space (Lawson, 2014). The way in which experts and expertise is understood in mainstream development is fundamentally ethnocentric, creating discursive associations with the North as advanced, progressive, and modern. The South, in contrast is constructed as "backward, degenerate and primitive" (C. McEwan., 2009). A patronizing 'will to improve', backgrounds claims of benevolence and neutrality embedded in technical and universal expertise. But expertise is more about dividing than unifying: "the practice of 'rendering technical' confirms expertise and constitutes the boundary between those who are positioned as trustees and the capacity to diagnose deficiencies in others, and those who are subject to expert direction" (Li, 2007, p. 7).

To expose the social and political roots of such constructions, development anthropologists put development theory and practice at the center of their own research, turning the inquiring gaze, the 'microscopes' back onto development experts and the discourses, practices, and power relations that were often taken for granted. Development, they argued, was itself a lens through which the world, and the world's communities, are constructed. The anthropology of development engaged more critical engagement "with the processes of directed change, especially with the way in which development discourse . . . may construct the object of development" (Grillo & Stirrat, 1997, 2020: preface).

Conclusion

Reflection and critique in development studies have fostered a rich history of evolving perspectives on development's experts over the past eight decades. These perspectives represent much more than our relationship with expertise, including the very definition of progress and development, and how we confront and understand difference and 'otherness'. It is apparent that many development practitioners and scholars have not yet fully done away with notions of the benevolent expert, but this imagery came under intense scrutiny since as early as the mid-1960s; critical discourses widened in the 1970s, and entered mainstream development discourse in the late 1980s and 1990s.

The hubris of expertise persists. The environmental crisis has inspired a new generation of techno-optimism and 'bright green' visions of the future that depend on high tech solutions to global environmental problems and poverty. To some extent, however, the current environmental and social crises have inspired a new generation of collaborative problem solving and interdisciplinarity, engendering a more critical and dynamic view of expertise as relative and contextual. As STEM students outside of development studies engage with the global problems of poverty and inequality, there is an increasing appetite for service learning and social justice. Without the benefit of an understanding of the roles experts, including technical experts and engineers, have played in this history and the trajectories of national and subnational development, there is a risk of repeating a doomed history. The relationship between experts and development reflects the very foundations of development itself: what is the good life, and what signifies progress toward it? Whose knowledge is valuable, and how do we know it is valuable? Understanding how divergent views on experts and expertise and their ethical/empirical underpinnings have responded to such questions is crucial to development studies generally, and for anyone working in and studying development who aspires to critical and reflective practice.

Part II

The next two chapters draw on design and transdisciplinary literatures to reframe – or decenter – science and technology in development as a central concept for development engineering. Chapter 3 focuses on the transdisciplinary nature of design thinking and practice. Derren Rosbach, Yunus Telliel, and Robert Krueger show that many transdisciplinary areas of research that are similar to development engineering have matured in their capability to integrate multiple disciplines under a coherent framework. For them, the biggest challenge for transdisciplinarity is to appreciate and incorporate nonacademic, experiential, and indigenous knowledges. They suggest that moving beyond the expert-nonexpert dichotomy, as discussed by Elgert in Chapter 1, has been more difficult as this requires a more radical rearrangement of knowledge production. Within this genealogy, they show how non-Western visions of technology design have been silenced. Chapter 4 takes this inquiry one step further by exploring the ethical frameworks for a design perspective. Whether it is policy, program development and implementation, technological development or innovation, the design process can help to reveal opportunities for empathic and equitable interventions. In this chapter, Telliel argues that because design is always about 'what could be', it has the potential to move away from the status quo of 'what is'. For this reason, he suggests, design is an inquiry into future possibilities, and designing for just futures is an ethical duty for engineers in the development engineering space. The ideas introduced here will be picked up in policy, technology, social science, and artificial intelligence contexts.

https://doi.org/10.1515/9783112218716-003

Derren Rosbach, Yunus Doğan Telliel, and Robert Krueger

Chapter 3 Transdisciplinary Thinking and Doing in Science, Engineering, and Development

Introduction

In the first two chapters of this volume, we initiated a discussion on some key conceptual and methodological challenges with development. Our purpose here was to set up this chapter, which will focus on ways of employing scientific and engineering knowledge and design in ways that promote knowledge egalitarianism and reconciliation. In this section we seek out ways to bring more people and perspectives into the engineering design process so that it can look not merely at a problem and a solution, but look forward and backward in time and across cosmologies.

To date, we remain disappointed in what has been called 'transdisciplinary research'. As you will read below, it has been a narrow enterprise that remains exclusive. In this chapter we examine the roots of transdisciplinary research as we know it. We then lay a foundation for what transdisciplinary research and design could be, e.g., more inclusive and egalitarian. Our first step in thinking about transdisciplinarity is to incorporate the concept of design. Here we privilege the design approaches to knowing and creating over traditional social science methods. Traditional so-called transdisciplinary research, while having its benefits to be sure, reproduces the subject-object dichotomy between the researcher and the researched. It also upholds the expert/lay dualism. This must be overcome if we are to create true partnerships in development engineering. We also want to push development engineers, researchers, and technicians to redouble their efforts to be responsible transdisciplinary cocreators.

Specifically, this chapter calls for the integration of transdisciplinary thinking (TD) into design as practiced by scientists and engineering in low- and middle-income countries. Engineering expertise has long dominated sustainable development work. Many efforts to confront development challenges at the community level are led by outsiders that eschew local knowledge and experience in favor of science and engineering expertise. Many of these efforts fail in a large part because of a lack of integration between scientific and local knowledge. Here we propose a transdisciplinary approach with a focus communication that employs local and indigenous knowledge and expertise as a starting point for a collaborative process that benefits from a grounding in practical place-based knowledge combined with the power of science and engineering.

https://doi.org/10.1515/9783112218716-004

A Brief History of Transdisciplinarity as a Concept and Practice

The term transdisciplinarity emerged in the works of Jean Piaget, Edgar Morin, and Erich Jantsch in the 1970s. They developed the concept of transdisciplinarity to encourage scholars to bridge disciplinary boundaries in a way that exceeded multidisciplinary and interdisciplinary approaches (Nicolescu 2002). The need for this approach came at a time when researchers became increasingly aware of (1) the complexity of emergent social, economic, and environmental challenges and (2) that disciplinary research specializations had led to a limited ability to generate novel solutions to such complexities (Klein, Grossenbacher-Mansuy et al. 2001). These initial efforts sought to transcend the limitations of traditional disciplinary practices of knowledge production and their subsequent applications.

During the 1980s and 1990s, various environmental risks and threats to human health brought scientists and social groups together in new ways (Clark 2003; Gibbons et al. 1994). Today, calls for transdisciplinary research have intensified in the context of framing and seeking solutions to 'wicked' problems such as climate change, energy transitions, and poverty – all of which are complex socio-technical issues facing our world that are too challenging for a single discipline or stakeholder group to solve on their own (European Commission 2012; Owen and Cowell 2011).

These issues require an increased effort to collaborate across disciplines and incorporate the diversity of creative and innovative capabilities of communities of experts and nonexperts. Stokols (2006) describes the core of transdisciplinary research as "the integrative quality and scope of transdisciplinary research products (e.g., hypotheses, theories) that set them apart from the more traditional intellectual products of unidisciplinary science". At the same time transdisciplinarity is also about building a space for creativity and innovation in problem analysis and solving. Integrative and imaginative nature of transdisciplinarity allows for scientific and technological knowledge to be circulated across and applied into multiple settings (Somerville and Rapport 2000).

Transdisciplinary approaches have been applied to a number of diverse concerns such as public health, sustainability, urban planning, and international development to name a few (Albrecht, Freeman et al. 1998; Balsiger 2004; Hadorn, Hoffmann et al. 2008; Stokols, Hall et al. 2008; Stokols, Misra et al. 2008). These examples generally display a common set of characteristics that distinguish transdisciplinary approaches from interdisciplinary and multidisciplinary approaches, but some ambiguity still exists about these. In particular, the most defining process of a transdisciplinary approach, the creation of a shared conceptual framework among researchers, NGOs, and communities, has not been fully developed in the literature. This becomes even more complicated if we think about transdisciplinary design in non-western contexts.

Multi-inter-transdisciplinarity

Categorizing transdisciplinarity is challenging for several reasons. Neither ideas nor terminology are not developed in a comprehensive way and many times are used interchangeably. The open-endedness of TD inquiries have made it difficult for TD practitioners and scholars to create platforms that enable 'smooth' translations between epistemological assumptions (e.g., what constitutes a legitimate object of inquiry from a discipline?) and methodological concerns (e.g., what do we do to learn about this object of inquiry?) Indeed, straying from the safety of disciplinary traditions has its perils, one of which is increased ambiguity as knowledge practices are dislodged from their deep disciplinary roots. This of course can also be a strength whereby disciplinary constraints are shed resulting in new questions and insights and – eventually – enabling the construction of new knowledge systems (Gibbons, Limoges et al. 1994). Further, hybridization between disciplines can lead to novel solutions to seemingly intractable problems (Stokols 2006).

Transdisciplinarity is only one way of organizing cross-disciplinary work. Rosenfield (1992, p. 1351) provided a useful and commonly cited classification based on three types: multidisciplinary, interdisciplinary, and transdisciplinary:

- Multidisciplinarity where researchers work from within their discipline on a common problem in either a sequential or parallel fashion. This is a typical practice in building construction, for example. Each practitioner participates in constructing one building, but works solidly from their discipline or profession as they do so. The architect conceptualizes the design of the structure and turns it over to an engineer or building construction manager, who then applies his or her disciplinary knowledge to move the conceptualized design toward a material manifestation. The various subgroups of builders (plumbers, carpenters, roofers, etc.) create the material structure. These specialists work from a common plan and share information, but each subprocess of production is segmented and conducted in a monodisciplinary fashion. Electricians design and install electrical wiring and systems, plumbers do likewise for water and waste systems, and so on.
- Interdisciplinarity occurs when participants work primarily within the confines of their disciplinary perspective, but also jointly on a common problem with members of other disciplines. example, an architect, engineer, and numerous building construction professionals might work together to design and build a suitable living space for a client. An interdisciplinary approach can lead to new insights due to a more open flow of information and cooperation, but each representative of the engaged disciplines still applies their individual expertise to the building process, and retains generated knowledge within their own discipline (Rosenfield 1992).
- Transdisciplinarity, however, suggests that researchers generate a shared conceptual framework or some form of meta-level coherence that draws from each par-

> ticipant's perspective, but is strengthened through the integration of disciplinary knowledge (Rosenfield 1992; Sage 2000; Stokols, Fuqua et al. 2003). Problem-solving that proceeds in this way results not only in a unique process, but also promotes novel outcomes.

To differentiate further: A multidisciplinary study is a collaboration among experts, members of different disciplines, where the relation among them is associative, i.e., where the work of each of them is added to that of all the others. These are indeed very common in many fields. In an interdisciplinary study the connection is relational. The disciplines collaborate in such a way that each takes up some of the assumptions and worldviews and languages of the others. A transdisciplinary study would therefore exist where the integrating relationship is taken to the extent of there emerging a transcendent metalanguage (McDonell 2000). Transdisciplinarity necessitates that contributors generate a shared conceptual framework, or some form of meta-level coherence that draws from each participant's perspective, to integrate relevant disciplinary knowledge (Rosenfield 1992; Sage 2000; Stokols, Fuqua et al. 2003). Problem-solving that proceeds in this way not only results in a unique process, but also promotes novel outcomes. Has argued that transdisciplinarity involves the creation of a transcendent metalanguage that is expressed by the various individuals engaged. Many analysts have adopted Rosenfield's (1992) argument that a shared conceptual framework is a definitive characteristic of transdisciplinary research (Parkes, Bienen et al. 2005; Stokols, Hall et al. 2008). In this view, TD is a process through which a common analytical perspective is generated, one that would not otherwise exist. Rosenfield (1992, p. 1351) has identified a conceptual frame that transcends "separate conceptual, theoretical, and methodological orientations in order to develop a shared approach to the research" as a distinguishing feature that differentiates interdisciplinary and transdisciplinary research.

From Shared Framework to Knowledge Creation

Transdisciplinarity can be viewed as a process through which different disciplinary perspectives, much like sensory perceptions, are integrated toward coordinated thought and action (cf. Rosenfeld 1992). This is similar to individual cognition when the richness of information processed by the human brain results in action. Individuals make innumerable choices as they process sensory information that are then compared to memories of experiences that identify which stimuli to notice, which to ignore, which require immediate action and which to store for later use. Disciplinary science has trained researchers to be very adept at collecting and filtering particular types of information using specific tools and techniques that are potentially very powerful, but at the same time can limit knowledge diffusion across disciplines. The potential of TD is not just to increase the breadth of problem solutions, but also to apply a diversity of perspectives to create a shared aspiration around what to look for, how to look for it, and why even to investigate in the first place.

One major challenge regarding shared aspiration and understanding is communication among individuals possessing different ways of knowing generated through diverse systems of representation. In a group characterized by disciplinary diversity, it is almost as if one person can only hear, one can only see, one can only smell, and another can only touch. How do they communicate each sensation to one another if each has never experienced the sensations others are describing, let alone the differing interpretations being supplied to describe those perceptions? The result is a gap in understanding and a potential tension among researchers (Hall, Stevens et al., 2005).

Another way of approaching knowledge exchange is to look more closely at the process of knowledge creation:

> Knowledge creation can be understood as a continuous process through which one overcomes the individual boundaries and constraints imposed by information and past learning by acquiring a new context, a new view of the world and new knowledge. . . . By interacting and sharing tacit and explicit knowledge with others, the individual enhances the capacity to define a situation or problem, and apply his or her knowledge so as to act and specifically solve the problem. (Nonaka, von Krogh et al. 2006, p. 1182)

These observations highlight the transformational potential of knowledge in a social context. Knowledge capacity is both limited and enhanced through relationships between people, information, and institutions of knowledge production. Latour (2005) has described the multiple relationships through which we access the world and accomplish work as networks of association. It is through these ties that experience and information are given meaning and new knowledge is generated. Knowledge is therefore intimately contextual. This is not to say that knowledge is bound simply by local experience. Rather, creation and application are contextual phenomena, and extraction and generalization of many forms of knowledge are challenging and limiting. Further, knowledge is transferred through relationships in the form of language or symbols that are representations of actual things or phenomena that only partially capture the full complexity of knowledge, especially tacit knowledge. Social justification of these representations is a fragile process that relies on leveraging multiple relationships (Nonaka, von Krogh et al. 2006). Sharing or transferring tacit knowledge through these relationships is a complex challenge, as Hildreth and Kimble have explained:

> Although under certain circumstances tacit knowledge can become explicit, it is perhaps more accurate to say that even then only part of what is tacit is made explicit. This is because even what we normally think of as explicit knowledge has a whole history of culture, conventions of language and cross-referencing of thought that are never made explicit. There is always more that is embedded, implied, assumed and presupposed than can ever be externalised and made

> explicit. In practice, the tacit and explicit dimensions to knowledge are inexorably and inextricably interwoven. (2002, para. 28)

This integrated duality of knowledge exposes a limitation of external aspects of knowledge creation, since much of what is known is tacit knowledge and is therefore difficult to express or transfer, and remains with individuals. Cognitive skills are tied to technology and social processes that are often hidden or normalized (Latour 2005). Latour has suggested social phenomena, normally hidden, become visible during technological or social[1] – or sociotechnical – disruptions. It is so because such disruptions (computer program stops working, an injury requires alternative means of commuting, etc.) cut the access to the know-how or tacit knowledge embedded in those ties or sociotechnical systems. As such, what Latour calls "networks of association" become visible when they fail to produce the anticipated result.

A simple example is when a computer hard drive crashes. The usefulness of the machine becomes painfully apparent when the information stored within it is no longer accessible. At the same time, processes supported by the computer become difficult, if not impossible. Even turning to another machine is only a partial solution since an individual's computer system evolves over time to support daily routines. Email addresses, gone; shortcuts on desktop, gone; customized application and programs, gone. As Clark has observed, when a hard drive crashes it feels a bit like a mild form of brain damage (2003, p. 4). Cognitive capacities embedded in a computer system (expanded exponentially through the Internet) are overlooked on a day-to-day basis, but become apparent during a disturbance. This is also true of people in organizations. Tacit and explicit knowledge are accessed and generated through relationships with key individuals. If an individual is no longer available, the extent to which they were relied on to accomplish work becomes very visible. Disruptions expose daily routines and capacities that normally hum along quietly in the background. However, exposing these quotidian processes can be a double-edged sword. Specifically in the context of knowledge production, exposing tensions or challenging norms can lead to conflict, but it may also lead to innovation (Stokols 2006). In one sense, a group becomes more self-aware when these knowledge creation processes are revealed and can then be adapted to the concerns or problem at hand. However, these knowledge systems or networks must be enacted continuously, and relationships must be maintained, if a system is to persist (Latour 2005).

Design as Practice-Oriented Transdisciplinary Research

In recent years, transdisciplinary research has evolved to incorporate stakeholder preferences as a guiding principle for research activities. This signals TD's shift from a research orientation toward a practice-oriented activity. For instance, the National Science Foundation's new initiative 'convergence accelerator' was launched in 2019 to support transdisciplinary science that is 'use-inspired and application-oriented' and 'integrates teams from industry, academia, nonprofits, government and other communities of practice'. We find this shift to be productive for new concerns and debates on transdisciplinarity. While being generative of new scientific understandings, the 'research' paradigm of transdisciplinarity has been limited in terms of what it is able to register as proper 'knowledge practices'. Indeed, for a long-time transdisciplinary research has valorized expert knowledge as, in the US, it has largely revolved around bringing together traditional disciplines, like biology and data science, into new disciplines, like bioinformatics. As Polk (2014) suggests, problem-solving efforts that aim to capture and address the needs and complexity of societal problems need to pay attention to contextual concerns of applied science and democratic norms of public participation than do traditional disciplinary approaches.

This is where we see the convergence between TD and design for science and technology for development. Design – both a process of making and an inquiry into that process – is itself constituted by transdisciplinarity. This is neither just because different teams of actor from different disciplines to work collaboratively to create new integrative conceptual, theoretical, methodological, and translational innovations, or nor just because design moves beyond discipline-specific approaches, singly focused forms of knowledge, and experts, to address socio technical problems. The 'design' paradigm of transdisciplinarity offers an opportunity to appreciate and engage with the multiplicity of knowledge practices that exist in a specific setting, as well as the contingency of social, technological, and environmental situations. Furthermore, it views the multiplicity of (academic and nonacademic) knowledge practices through the lens of contingency. Through the lens of contingency, what seems a failure in the integration of different knowledge systems is an opportunity to reenvision a new path. Design, as a process of co-making, allows a transdisciplinary project to incorporate failures as 'necessary' steps toward the success of the overall project. Furthermore, design as an inquiry into the very process of co-making directs transdisciplinarity from the epistemological to the ontological – different understandings of the world we inhabit. Because of its openness to diverse 'ontological' or 'cosmological' perspectives, design is as concerned with not-fully-articulated beliefs, values, and knowledges – 'stuff' of everyday life' – as it is with fully-articulated knowledge systems. As such, trandisciplinarity of design makes collaborative processes the very condition of possibility of a project, not simply a criterion of success.

It is true that many transdisciplinary project would approach design one discipline among many. Here we are thinking of – theorizing – design as a meta-discipline of problem-solving. Peter Buchanan (1992), as one of the figures that advocated for design's general relevancy to those who are interested in problem-solving before 'design thinking' became popular, argued that wickedness of problems in design is characterized by a fundamental indeterminacy. If design is to "conceive and plan what does not yet exist", according to Buchanan, designers think of design in a general or particular level. On a general level, designers reflect on the nature of the 'designed', the 'artificial', and the scope of its application. While such reflections can be the basis of a framework that guides the designer, they cannot translate into a determinate science. This is so, Buchanan suggests, "design is fundamentally concerned with the particular, and there is no science of the particular" (p. 17). On a particular level, designer deals with "specific possibilities of a concrete situation". Yet, such possibilities are not about "undetermined subject[s] waiting to be made determinate", but "indeterminate subject[s] waiting to be made specific and concrete". Buchanan suggests that designers do not need to turn design into a subfield of arts or sciences – not a discipline like any other. What designers need to determine how knowledge from the arts and the sciences "may be useful to design thinking in a particular circumstance without immediately reducing design to one or another of these disciplines". He calls this designer's "principle of relevance" which he believes should guide designers as they identify and incorporate all knowledge needed to complete a project.

Buchanan wrote at a time when design transitioned from simply being seen as making objects or improving the customer's experience, to restructuring the conditions of human and nonhuman life. Indeed, the outputs of contemporary design are increasingly diverse. As such, the traditional idea of design – as a single discipline – can not help capture the multiplicity of the processes and practices involved in the making of stuff. Some design theorists such as Elio Manzini suggest that this is a reflection of changing social dynamics. Manzini, for instance, argues that if industrial society centered on product design and late industrial society centered on service design, contemporary societies operate through networks that contain products, services, and systems. In a network society, from Manzini's perspective, a service economy comes together with a high degree of connectivity, and, as such, design thinking moves from 'value chains' to 'value constellations'. The impact of design in a network society is thus expansive. In reflecting on the relevance and meaning of design in a network society, Manzini advocates for a transformational perspective on design: "People's behavior cannot be designed. However, it is possible to create conditions that make some ways of being and doing things more probable than others" (p. 152). Manzini calls this 'infrastructuring' – design as the creation of favorable conditions for certain social and ecological transformations.

This shift is also reflected in the conception of the relationship between technology and society. While designers traditionally aim to translate a new technological idea into socially acceptable products and services, in this new framework designers are mainly expected to conceive and develop 'appropriate technologies': that is, products, services, or systems that are specifically designed in response to a social need. The latter puts technology as an extension of society while the first approaches technology as a separate realm of expertise. Unsurprisingly, the move to system-level awareness in design theory has accompanied a greater interest in societal transitions among designers. Indeed, we are living in 'transitional times'. From Ezio Manzini, to

Terry Irwin and CMU Design School, or Damien White, many design theorists argue a meaningful transition to sustainable futures can happen only as unsustainable institutions and systems are un-designed while new ones are designed. What we see in contemporary design scholarship a sea-level change is a recognition of design as a meta-discipline that is dedicated to framing and solving wicked problems. If transdisciplinary is not a single methodology and instead a meta-methodology for the practice of science and engineering, design – at least how it is defined in contemporary theory and practice – is a meta-discipline that serves as the vehicle for the practice of transdisciplinarity.

It is true that there have been numerous attempts to leverage design as an agent for positive social change, but very few have articulated how this change would be catalyzed across interconnected scales (in ecosystems, economies, organizations, communities, individuals, etc.). This is where transdisciplinarity is essential. It provides three things:

First, theoretical lenses. A wicked problem is essentially a transdisciplinary design concept. As such, a problem appears as a wicked problem only if seen through the lenses of a transdisciplinary designer.

Second, aligning inter-scalar connections in societal transition. How does a neighborhood connect to a broader ecosystem concern regarding flooding? Although not all multi-scalar problems are 'wicked problems', all 'wicked problems' are multi-scalar and 'wicked' in the sense that they defy a definitive formula or have multiple explanations or solutions that are contingent on the attitude of the researcher or designer. Transdisciplinarity allows for a way of aligning multiple processes and practices at different scales towards a shared concern or goal.

Third, transdisciplinarity addresses the whole human: from economic interests to the need for spiritual connection to concerns regarding public spaces. A transdisciplinary framework can facilitate collaborations among social scientists, humanists, artists, professional designers, and engineers and other technologists.

The Dar Si Hmad Community Fogwater Project

One of the advantages of talking about transdisciplinarity in these ways, is that it moves the questions of transdisciplinarity away from individual designer's intentions and responsibility to the needs of societal transitions. To explore the concept of transdisciplinary design and problem-solving we turn to a project from Southern Morocco, the Dar Si Hmad (DSH) fogwater project.

DSH for Development's fog-water project is an interesting and illustrative case of the potential of transdisciplinary design for development. Founded in 2010, DSH for Development, Education and Culture is an independent, nonprofit organization in

Southwest Morocco. They offer educational experiences to Moroccan and foreign students, as well as local students through their 'water school'.

DSH promotes local culture and creates sustainable initiatives through education and the integration of scientific ingenuity with the communities of Southwest Morocco (Dar Si Hmad, n.d.). Fog water catchment systems have been put in place by DSH to provide potable drinking water to several villages at the base of the mountain. A cell-phone-based communication system that is accessible for illiterate women was also developed (Dodson, 2014).

The Mount Boutmezguida fog water project is DSH's most celebrated initiative to date. The fog-water harvesting site sits on top of Mount Boutmezguida, located in the Aït Baâmrane territory of Southwest Morocco. In this semi-arid region between the Atlantic Ocean and the Sahara Desert, residents had been carrying water from wells. But since March 2014, water harvested by the fog-collecting nets is the main source of water to households, serving approximately 400 residents.

In 2006, Dr. Jamila Bargach, an anthropologist by training and professor at the University of Mohammed V, recognized the potential for using fog water capture technology as a sustainable water practice in a Berber community that was seeing its ground water diminishing. Evaluated the potential for fog water harvesting in the region and realized that Mount Boutmezguida's microclimate produced fog as the ocean air rises up the steep coastline to the crest of the mountain. Using ingenuity, perseverance, and raw muscle, Bargach brought the first nets to the remote site in 2013.

A number of aspects of this case stand out as illustrative of transdisciplinary research and design. First was the acknowledgement that increasing water scarcity was more than a water problem. The implications of diminishing a water supply had detrimental effects beyond potable water and agricultural applications. Also at stake were girls' access to education, community health, and women's livelihoods. In particular, women spent an average of 3.5 h every day fetching water from open wells. This laborious, time-consuming work was considered a woman's role in nearly every household (Dar Si Hmad, n.d.). Water-hauling and other domestic chores prevented many young girls in the area from attending school.

This early recognition of the social concerns around addressing water scarcity required the participation and collaboration with the local community in order to effectively design fog water capture technology that would address local conditions and community needs (Marzol and Megia, 2008). As a result, challenges with the initial fog water capture nets led to subsequent designs that addressed issues encountered by users and led to a design that required minimal simple maintenance. Also, through community consultation the designers and contractors addressed questions of water access, distribution, and delivery. People wanted water delivered directly to their homes, not a central location. This required a complex system of piping, valves, and meters. Therefore, a monitoring and maintenance regime also needed to be developed. These activities involved a variety of jobs, including installing and maintaining the nets with DSH staff, taking care of water valves, reporting any problems occurred to the DSH staff, etc.

Because of the new technology, women who once had the important position of being the purveyors of water, were supplanted by men as the caretakers of the water supply. This is a result of Berber culture where women can only use certain kinds of technology. For example, when mechanization was introduced in the region, women were given the donkeys that were rendered obsolete for farming and transport. However, women could not use these new technologies themselves (Dodson 2014).

The fog water distribution system was no exception. It presented problems because of the society-technology relationship.

- Women, who were the primary gatherers and users of water for cooking, cleaning, and for animals had lost their connection to the resource because wells and boreholes with hand pumps were obsolete. The networks of pipes, meters, and other technology, excluded women from managing the resource except for inside the home.
- The fog water intervention also increased the isolation of women from their community because they no longer had to fetch water; there was no reason to leave the compounds where they live. They lost their ability to socialize with other women (they can't talk to men) and share news about the community.
- Finally, if there was a problem, they could not report it to a repairman because they cannot speak to men who aren't their relations. This caused problems with metering and waste.

For many organizations and donors, delivering a sustainable source of water would have been the end of the story. Project complete. Success. Let's move on. This was not the path that Dar Si Hmad took. Because they are embedded in these communities, they understand the complications that the fog water collection and delivery system brought to the communities, particularly women.

Bargach, working with Leslie Dodson, a PhD student at the time, developed a technical intervention that gave back women some of their status. And, working with the Dirham Foundation, they created social and economic opportunities that were culturally appropriate. As part of her PhD research in global engineering, Dodson worked to develop mobile technology using function phones. Women are not allowed smartphones so a 'function-phone' had to suffice. Dodson's system made it possible for all women, literate or illiterate, to send SMSs when there was a problem with the system. This system provided one opportunity for women to overcome communication constraints that women face in the region and to ensure that the community continues to benefit from access to a reliable source of clean water (Dodson, 2014).

In addition to these interventions, Dar Si Hmad helped to establish argan nut women's cooperatives. Through these opportunities women had their social lives returned to them as the crushed argan nuts for use in beauty products in the West. It also gave them money. When a woman earns money in Berber culture it is hers to use after the household bills are paid. When a surplus is available women can pay to send their daughters to school beyond the sixth grade.

Dar Si Hmad's fog-water harvesting project resulted in the largest fogwater system in the world and represents the collective efforts of a diverse set of actors including a number of NGOs, social scientists, climate scientists, engineers, and community members (Dodson and Bargach, 2015). Together, they managed to design and implement a functional system that incorporated local knowledge, engineering expertise, and community engagement to relieve water stress in a rural Berber community (Dodson and Bargach, 2015).

Conclusion

This chapter has focused on ways that theorists, scholars, and practitioners of development can rethink the design process for development. Elaborating on the different approaches to disciplinarity, multidisciplinarity, interdisciplinarity, and transdisciplinary demonstrates for us what is key in understanding thoughtful approaches to science and technology for development. For us, the task is to learn to "grasp the other's position in the strongest possible light . . . not as an adversary, but as a conversational partner". We argue that to overcome the ideological and power asymmetry we call on the approach that what Eglash (2007) has referred to as "peer-to-peer engagement". In real, peer-to-peer engagement one form of knowledge is not seen as an ultimate arbiter or universal creator of solutions. Rather, it requires the (very) hard work of understanding the wide range of perspectives that are attendant to a particular problem.

Taking additional lead from geographers Barnes and Sheppard (2010, p. 194), who draw on the concept of engaged pluralism, we argue that the process should involve resolving that even though "we are committed to our styles of thinking, we are willing to listen to others without suppressing the otherness of the other". We must, therefore, celebrate knowledge diversity rather than use 'diversity' as a way to reinforce false binaries and scientism.

Those following an object oriented ontology or speculative realist approach understand design and outcomes are related to process, complexity, and contextual relations. Scholars such as eschew the search for absolutes because it only leads to the subjugation of 'other' forms of thinking. Yet, this kind of thinking is what transdisciplinarity seeks to transcend. To riff off of John Dewey's 'spectator theory of knowledge', which relies on the metaphor of vision as the primary model for knowing, we only see what we are trained to see or what our norms allow us to see.

History is fraught with such behavior. The literature on imperialism, colonial expansion, and Science and Technology Studies describes how people in sub-Saharan Africa, for example, used the 'scientific method' prior to its discovery in Europe.

Eglash's example of fractals in West Africa is an excellent example. Indeed, people of Senegal, Burkina Faso, and Ghana, for example, created fractal shapes hundreds of years before fractal geometry was discovered in Germany or found useful in US computing – through interdisciplinary engagement, by the way – in the 1960s.

Approaches to science and technology for development has traditionally reflected the spectator theory of knowledge in what has often been the blood sport of development. For example, engineers 'see' a problem: the lack of economic opportunity, the need for a bridge, or electrification.

From a transdisciplinary perspective there is a whole lot more out there to see than an infrastructure problem. 'Problems', and especially development ones, have pasts and futures. As we point out in the next chapter, rural economic development in East Pakistan and Uruguay, created the material accoutrements for development, but did not envision why these problems were perceived in the first place or the implications fixing said problem would have on the groups they sought to help. Bringing thoughtful, challenging, and at times maddening, collaborations is a way to see across the different components of problem identification and problem-solving. For us, design emerges from social, ecological, scientific and technological interactions, therefore lab-developed technology, e.g., building, water, and energy systems, must be informed by and responsive to those concerns (Dunne and Raby 2013; Malpass 2018).

Yunus Doğan Telliel

Chapter 4
Design as Inquiry: Ethically Aligned Design and Matters of Justice

Introduction

Critical design theorists have suggested that we live in a world in which almost everything we use and interact in our everyday life (from computers to chairs to built environments) are designed. Historically, in development, these can be bridges, canals, railroads, dams, or factories. Development also includes social practices, such as the practice of design.

Where the previous chapters covered topics on expertise in its multiple forms, this chapter explores design's ethical premise: its ability to undesign and redesign engineered systems so that sustainable futures for the planet and its denizens can emerge. Defuturing is the term Tony Fry (1999) coined to capture modern design projects' erasure of sustainable futures. If defuturing is the process of depleting resources and damaging ecosystems to the degree that future generations will be robbed off of the possibility of pursuing collective wellbeing, futuring is the reversal of those forces and conditions of defuturing. In any design process, the designed carries the yet-unrealized potentiality of being designed otherwise and the possibility of not being designed at all. Inspired by Fry's framework, I propose design as inquiry as a framework to think about the role and place of temporal critique in engineering design. Conceptualized as an inquiry, design is not simply about making and creating new objects, services, or systems, but also a way of investigating the possibilities of alternative futures that design can bring into existence.

All technologies are created through design – imagining and making something new out of what already exists. In most technology development projects, engineers focus on what is possible within the reach of technical possibilities. Their question tends to be less about 'should we do it just because we can?'. Without such a question, technology design seems to be a straight-forward realization of what is already within the reach of designers' conceptualization. Thinking about design as inquiry helps recognize a tension among 'what is', 'what could be', and 'what ought to be'. As the latter complicates straight lines drawn between 'what is' and 'what could be', design becomes an inquiry without a predetermined answer. It invites the engineer to be an ethical participant in the creation of technologies that serve the common good.

Many scholars have criticized techno-solutionism as a major problem in technology research and development (Morozov, 2013; Levina and Hasinoff, 2017). Defined as the reductionist interpretation of complex social issues as problems that can be fixed by technical solutions, techno-solutionism is what left off of design when ethical in-

https://doi.org/10.1515/9783112218716-005

quiry is eschewed. Without a consideration of future possibilities of 'what ought to be', 'what could be' becomes the technological fix that registers neither cultural differences nor social injustices. The temporality of the designed does not only extend into the future, but also connects the present to the social structures and ideologies transmitted from the past. As many predominant values and preconceptions are inscribed into the designed, the designed becomes the vehicle that reproduces the power relations of the past in the present. Facial recognition systems, so-called predictive policing algorithms, workplace surveillance tools, and sex robots are all recent technologies that have given new life to past thoughts and practices that have embodied ideologies of capitalism, racism, and sexism. In design as inquiry, 'what ought to be' is the disruption of the very process of reproduction of such ideologies in new technologies.

Of course, not everyone sees design's concern with 'what ought to be' as an ethical imperative. In technology companies which are led by profit-making more than anything else, engineers are often disempowered to even think about 'what ought to be'. What is expected from them tends to be that they move from 'what is' to 'what could be' as fast and less costly as possible. Sometimes engineers might find themselves in situations of what Andrew Jameton (1984) called moral distress: knowing what to do to be ethical, but are not be allowed to do it. In academic research, a different type of incentive leads to a similar consequence. Even though the problem is not profit-making as in private sector, the prestige economy of research funding rewards the new over the sustainable, the accessible, and, ultimately, the ethical.

In this chapter, I argue that it is increasingly more important for engineers to keep 'what could be' and 'what ought to be' at a critical distance from each other. This distinction, I suggest, is one of the keys to revive an ethical imagination that can redirect technology design for the common good. This requires mobilizing design inquiries to create artifacts, interactions, or systems that can restructure the conditions of human and nonhuman life in ways that benefit them all, as well as can foster ethical relations among humans, nonhuman beings, and things. From this perspective, design brings forth a set of temporal horizons. It explores future possibilities with respect to past experiences and present potentialities. In order to imagine multiple ways in which a design could come into being, the engineer needs to navigate between distinct design options. This sometimes means deciding not to design, as sometimes nondesign is preferable because it means stopping a design project that can potentially bring harm to a community or a group of individuals. Design as inquiry is a way to capture design's ethical premise as reflected in an engineer's awareness of the possibility of designing otherwise and its own negation via refusing design.

The chapter is divided into three sections. I begin with a section on engineering design. I focus on IEEE's ethically aligned design framework, and discuss its value with respect to 'design as inquiry'. While the ethically aligned design framework has shaped a number of important initiatives including IEEE standardization projects, it tends to take for granted the distinction between technology and society. Even though this distinction allows the ethically aligned design framework to be easily adaptable to an engineering project, this framework accepts the need for a technology and tries to integrate ethics within an already predetermined trajectory of design. It does not encourage engineers to consider whether a particular technology is needed at all.

The second section of the chapter moves to a discussion of Bruno Latour's idea of design for 'matters of concern'. In his engagement with design scholarship, Latour suggested that design has the power to shift the public attention from 'matters of fact' to 'matters of concern'. I agree with Latour's suggestion that design can play a crucial role in mobilizing public attention to alternative socio-technical systems that can support our interconnected lives. 'Matters of concern' constitute a form of futuring guided by 'what ought to be'. But, there remains a vital question: whose 'concerns' are going to be heard – be listened to? In order to prevent the domination of some voices over others, I suggest that design for 'matters of concern' needs to keep its eye on 'matters of justice'. This is important because concerns for redesign can be easily used to exclude workers and other communities whose interests, voices, and visions are not well-represented in current power structures. With a focus on 'matters of justice', design as inquiry connects the conversation from what the future needs to who is part of this desired future.

I do not know if it is possible to unify voices that exist across cultural differences and social hierarchies. The third and last section focuses on this very question: to what extent is design able to negotiate and mediate across multiplicity of interests? Instead of imagining a movement that brings everyone together under one singular vision for redesign for sustainable futures, I offer a vision of design that is open to contestations and conflicts. While such a movement is perhaps not realistic, it also plays an ideological role by masking inequities and injustices. By acknowledging and valuing dissent, the principle of agonism can support open-ended inquiries about what the desired futures look like and how design gets us there. In this regard, the ethical power of design as inquiry ultimately comes from the ability to learn to imagine with others.

'Ethically Aligned Design' and the Impact Paradigm

Ethical design has been receiving a great deal of attention among engineers and other technologists. There is a growing awareness that codes of ethics, while being indispensable to professional engineering associations, are not often sufficient to fully address complex issues emerging in design practices (Warford, 2018). In parallel to this, some engineers move from a conception of ethics as something reflective of the individual engineers' intention and responsibility to a critical awareness of social injustices that engineering projects are (unintentionally or not) creating or furthering (Leydens and Lucena, 2017). With this shift, technology design is not simply about making artifacts or improving customer experience, but also about reshaping social conditions. Seen from this broader perspective, design comes to lie at the heart of most of the current debates on questions of social equality, cultural diversity, and environmental justice. In moving the focus from the individual designer to the broader social context, engineering ethics has more to say about the kinds of politics and communities that technology design practices can make possible.

The ethically aligned design report prepared by The IEEE Global Initiative on Ethics of Autonomous and Intelligent Systems is one example of this shift (EAD, 2021). IEEE is the world's largest engineering association. As of December 2022, it has over 427,000 members in more than 190 countries, more than 64% of whom are from outside the United States (IEEE, n.d.). As an industry-facing program under IEEE, the Global Initiative's mission is "[t]o ensure every stakeholder involved in the design and development of autonomous and intelligent systems is educated, trained, and empowered to prioritize ethical considerations so that these technologies are advanced for the benefit of humanity" (IEEE Standards, n.d.). The framework of the ethically aligned design report reflects this mission, as it provides a set of ethical guidelines aiming to embed ethics in engineering design. The ethically aligned design framework is built on five high-level guiding principles including human rights, well-being, accountability, transparency, and awareness of misuse. The 263-page long report also offers ethical guidelines on granular issues such as personal digital data, general economic benefit, legal policies, industry standards, and education and public awareness.

The ethically aligned design report joined a multitude of ethics documents focusing on big data and autonomous technologies that appeared since the late 2010s (Fjeld et al., 2020). The report represents a major achievement for IEEE. Indeed, it has led to a number of IEEE initiatives that aim to produce 'industry standards' that will guide technology companies that are interested in incorporating ethical design principles into their products and services. As of June 2023, out of fourteen standardization projects, eight of them are completed and the reports on each are available at the IEEE standards website (IEEE Standards, n.d.). These IEEE standards include areas such as 'process for addressing ethical concerns during system design', 'transparency of autonomous systems', ' data privacy process', 'transparent employer data governance', 'ethically driven robotics and automation systems', and 'assessing the impact of autonomous and intelligent systems on human well-being'.

IEEE's ethical design standards represent an institutionalized attempt to implement and promote an applied ethics framework in technology design. These standards are not meant to further scholarly conversation, but to ensure that "engineers, as the designers, companies as promoters, and authorities as regulators [are] aware that they all should go beyond technical issues and provide concrete assurance to the public that new developed [technological] systems are not harmful to the public in any ways and are for the human wellbeing" (Shahriari and Shahriari, 2017). For instance, IEEE standard for transparency of autonomous systems, one of the completed standardization projects, has two functions. First, the standard serves as a tool to assess the transparency of an autonomous system. Engineers and companies can use the standard and its measurable rubric to identify the level of the transparency of an autonomous engineering system needed for a range of stakeholder groups (i.e., users of autonomous systems, the general public and bystanders, validation and certification agencies and auditors, incident investigators, expert advisors in administrative actions or litigation). Of course, each stakeholder group has different expectations. Secondly, the standard provides a guide to the transparency measures that should be considered during technology design and assessment. The standard is not prescriptive in terms of how these transparency measures should be implemented. This gives a considerable amount of flexibility for how engineers and companies can approach the standardization process.

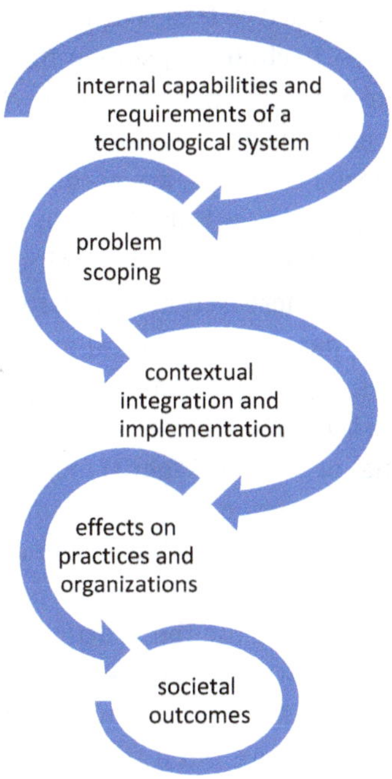

Figure 1: Impact paradigm of engineering ethics: moving from the **[technology]** station to the **[society]** station.

The IEEE standards that build on ethically aligned design framework often take a stakeholders approach in order to encourage engineers, companies, and regulators to think of major processes entailed in the development, deployment, use, maintenance, and failure of a technological systems. As such, they are mindful of potential contextual and pragmatic questions of its implementation. What I find interesting here is the power of design in bringing social and ethical concerns to the heart of engineering practices. This power of design is drawn from its dual nature. In the ethically aligned design framework, the implicit assumption is that design can never simply be about technical capabilities of a technology. 'The social' is always part of the life cycle of a technology from its inception to its maintenance. Design's dual nature (the technical and the social) opens up a space of ethical reflection on probable consequences of technological innovation and implementation. For this reason, the mutual evolution of understandings of ethics and design is at the heart of such IEEE initiatives.

While the ethically aligned design framework fosters a new way of thinking about ethics in engineering, it still operates within a paradigm that is primarily based on a model that starts from technology's impact on society: [technology] → impacting → [society]. This paradigm – what I call 'impact paradigm of engineering ethics' – separates technology and society as forces that can affect each other in potentially beneficial or harmful ways. Yet, technologies (and the supposed needs for specific technologies) are often taken for granted. The very possibility of non-designing certain technologies is rarely considered. It is true that the impact paradigm offers engineers and other technologists a practical methodology and a productive lens. Within this paradigm, they can start their ethical assessment from the [technology] station they are comfortable with, and work their way through from this station to the [society] station. Nontechnical knowledge, such as information about users, regulators, and legal frameworks, or awareness of environmental and social concerns, are gradually incorporated in the process. Such nontechnical knowledge helps with the breadth of problem scoping in engineering design, the kind of engagement with communities that may be impacted by such technologies, as well as the increased awareness of technological interfacing with humans. Furthermore, this paradigm is central to the National Science Foundation and other funding agencies supporting academic engineering research. For instance, one of the NSF's two merit review criteria is broader impacts of research. While the other criterion (intellectual or technical merit) focuses on the potential to advance knowledge, the broader impact criterion is for evaluating a proposed project's "potential to benefit society and contribute to the achievement of specific, desired societal outcomes" (NSF, n.d.).

Even though the impact paradigm continues to be valuable for the practice of professional engineering, the multiple systemic crises our world has been going through seem to demand alternative forms of engagement. My contention is that these crises do not necessarily ask for new technological solutions, but a framework of technology design that can guide socio-technical imagination and investigation of future possibilities – what I call design as inquiry. Take climate change, for instance. This is perhaps the first truly global crisis that poses challenges to conventional food production, infrastructure, public health, and many other systems. While every society – regardless of their wealth – has something to lose, the global response to climate change remains sluggish. Indeed, while the public discourse is dominated by to-do lists for mitigating the effect of climate change, conversations on its root causes and the beneficiaries of a global carbon-based economy often fall flat. It is as though that the problem is not that we do not know what to do, but that we lack a framework for global collective action. Climate change crisis thus entails a crisis in ethical imagination (e.g., an inability to imagine ourselves as part of relations of interdependency).

The impact paradigm of engineering ethics lacks the capacity to generate such ethical imagination. It recognizes design's power to connect the technical and the social, and puts 'the ethical' as a bridge between these two; that is where it stops. Radical ethical possibilities of design remain to be inquired.

From 'Matters of Fact' to 'Matters of Concern' – and to 'Matters of Justice'

In his address to the Design History Society conference in 2008, Bruno Latour (2008) argued that design emerges as a crucial lens of critical theory as "the typically modernist divide between materiality on the one hand and design on the other is slowly being dissolved away. The more objects are turned into things–that is, the more matters of facts are turned into matters of concern–the more they are rendered into objects of design through and through" (p. 2). For Latour, the view from design offers a new horizon to think about the world around us. As design gets the spotlight, Latour suggested, narratives of unilinear progress, detachment, and mastery are replaced with narratives of layered histories, attachment, and interdependence. He added that "the word design is a little tracer whose expansion could prove the depth to which we have stopped believing that we have been modern. In other words, the more we think of ourselves as designers, the less we think of ourselves as modernizers" (p. 3).

I am not sure if posing design against modernization will deliver the theoretical insights Latour sought to offer in his lecture. There are still dominant ways of thinking about design that are very much in line with those narratives of modernization. But, I agree that design has this type of potential, and the idea of design as inquiry focuses on it. 'Matters of concern' – not 'matters of fact – constitute a departure point for an inquiry since the latter is taken simply as data that is separate from political and social issues. "Instead of taking products or services as a starting point for design", Peter Friedrich Stephan (2015, p. 216) explains, design for 'matters of concern' can ask interesting questions such as "what concerns make people behave in a certain way? How do individuals interact with one another, with artefacts, and with the environment? Why do people speak and dress the way they do? Why do people buy the things they do?"

Let me focus a piece by Latour to show how 'design as inquiry' aligns with 'designing for matters of concern'. Early in the COVID-19 pandemic, reflecting on the disruptions in the working of capitalist production as a whole, Latour (2020) wrote:

> The first lesson the coronavirus has taught us is also the most astounding: we have actually proven that it is possible, in a few weeks, to put an economic system on hold everywhere in the world and at the same time, a system that we were told it was impossible to slow down or redirect. To every ecologist's argument about changing our ways of life, there was always the oppos-

> ing argument about the irreversible force of the 'train of progress' that nothing could derail 'because of globalisation', they would say. And yet it is precisely its globalised character that makes this infamous development so fragile, so likely to do the opposite and come to a screeching halt.

The COVID-19 pandemic turned into a major economic and social crisis very soon after it was recognized by medical authorities. The suspension of 'work' we know of it, the disruption of supply chains, and the impact on workers and their families all contributed to the questioning of some of the taken-for-granted aspects of modern life. What surprised Latour was that the systems and mechanisms of global capitalism were able to be suspended to respond to a global crisis of such destructive nature. Latour (ibid.) added, "[i]njustice is not just about the redistribution of the fruits of progress, but about the very manner in which the planet is made fruitful".

At the end of this piece, Latour (ibid.) invited the readers to reflect on "what [they] are attached to; what [they] are ready to give up; the chains [they] are ready to reconstruct and those that, in [their] behavior, [they] have decided to interrupt". He further offered a long list of questions:

- What are the activities now suspended that you would like to see not resumed?
- Describe why you think this activity is harmful/superfluous/dangerous/inconsistent and how its disappearance/suspension/substitution would make the activities you favor easier/more consistent.
- What measures do you recommend to ensure that the workers/employees/agents/entrepreneurs who will no longer be able to continue in the activities you are removing are helped in their transition toward other activities.
- Which of the now suspended activities would you like to develop/resume or even create from scratch?
- Describe why this activity seems positive to you and how it makes it easier/more harmonious/ consistent with other activities that you favor and helps to combat those that you consider unfavorable.
- What measures do you recommend to help workers/employees/agents/entrepreneurs acquire the capacities/means/income/instruments to take over/develop/create this favored activity?

For him, these questions could be a starting point for a collective exercise which can eventually lead to a radical reimagination of societies and environments. While this might be the first step, the generalization of such a reimagination, I argue, needs design as inquiry: Which designed systems are enabled by new activities? What designed systems should be undesigned? Which new systems should be designed to replace the ones that are undesigned?

Here, crisis (the COVID-19 pandemic in Latour's case) is an invitation to translate 'matters of fact' to 'matters of concern'. What does it mean to think about crises (climate crisis, public health crisis, etc.) through the perspective of design? Crisis is a rupture in the passage of time. It is the pressure of time that necessitates judgment. As

Reinhardt Koselleck (2000) argued, critique and crisis have been closely related in modern cultural discourses. Design can be the bridge that ties crisis and critique to ethical imagination. It can inform the judgment that will transform what is in crisis. Design can do this by facilitating new habits, practices, and even life forms as it proposes practical and meaningful alternatives. Seen in this way, design can be: (1) an inquiry of critique that reflects shared concerns and recognizes the conditions that gave rise to the crisis, and (2) an inquiry of making that materializes artifacts and interactions that embody shared concerns and undo those very conditions.

Maria Puig de la Bellacasa (2017) extends Latour's 'matters of concern' to bring attention to the aesthetic, affective, and ethical implications of concerns. de la Bellacasa's renaming of 'matters of concerns' as 'matters of care' provides a much-needed directionality to the idea of matters of concern. Of course, some concerns can be oppressive, silencing other concerns. Furthermore, the recognition that the world is populated with 'matters of concern' does not necessarily lead to a design that promotes ethical restructuring of social conditions and relations. With a shift to design for 'matters of care', there can now be a possibility to articulate an ethical imperative to see design from the perspective of interdependencies among humans and nonhumans. Inspired by de la Bellacasa's, I want to bring another concern to design: justice. Justice as a concern is essential to mobilize design as a critical inquiry into the designed that reproduces asymmetrical power relations in societies. Design for 'matters of justice' demand an act of undesigning to redesign. This is the ethical premise of design's self-negation, as well as the very condition of designing for a just society.

Whose Justice, Which Design?: Agonistic Design for Just Transitions

Design – as it focuses on 'matters of justice' – needs to attend to transitions and transformations. From electric vehicles to alternative energy infrastructure, there is a growing popular demand in shifting to social and economic processes that are not dependent on carbon-based systems. In parallel, design scholarship has been engaged in issues of climate change and sustainability, and articulated a range of strategies to enable societal transitions. Ezio Manzini's 'design for social innovation' (2015), Tony Fry's 'futuring' (2004) and design as 'redirective practice' (2007), Arturo Escobar's 'autonomous design' for the pluriverse (2018), Irwin, Kossoff and Tonkinwise's 'transition design' (2015), and Damien White's 'design politics for a new green deal' (2020) are all significant contributions to this line of investigation in design studies.

One of the central challenges for design researchers and practitioners is how to center social justice concerns (or what I call 'matters of justice') in societal and environmental transitions. This question is becoming urgent as the idea of sustainability transitions is now taken up by corporations and states. Advocating for sustainable futures is no more simply a matter of demands proposed by bottom-up movements of local activists. For instance, the United Nations Environment Programme (UNEP), a global authority on the environment, recognizes 'green economy' as one of the priorities areas. In collaboration with other United Nations agencies including the International Labour Organization, the UN Development Programme, the UN Industrial Development Organization, and the UN Institute for Training and Research, UNEP has recently introduced the Partnership for Action on Green Economy (PAGE), a platform that connects policymakers and sustainable development experts around the world. Furthermore, a 2023 report from Arup, a multinational sustainable design and development company, and Oxford Economics, an economic forecasting company, estimates that new green industries have the potential to be 5.2% ($10.3 T) of global GDP by 2050 (Oxford Economics, n.d.). It seems that the idea of an economy that does not rely on nonrenewable resources and unsustainable practices is now part of the global governance discourse.

As demands for sustainable futures are gaining popularity, an increasing number of corporations have been co-opting this language without necessarily changing their institutional policies and practices. Indeed, as researchers such as Laufer (2003) and Macellari et al. (2021) show, greenwashing (misrepresentation of an organization's actual commitment to the environment) and bluewashing (misrepresentation of actual commitment to the UN Sustainable Development Goals) are now common in corporate talk. So, the question is: how does transition design accomplish the goal of creating sustainable futures while working towards establishing policies and mechanisms that privilege the communities whose interests do not align with those of global capitalism and governance structures?

First of all, neither all these transitions are the same nor all benefit everyone. There are more unjust transitions than just ones. The sociologist and design theorist Damien White (2020) claims that the key for a just transition is the prioritization of "the struggles of displaced fossil fuel workers and communities", the centering of "interstate, interregional or intersectoral inequalities", and the advocacy of "intergeneration justice" (p. 6). In order to do accomplish this, according to White, public investment must be "directed towards high-quality, sustainable public housing and climate-smart infrastructure to meet pressing needs, ensure swift decarbonization and maximize urban resilience to future climate challenges" (p. 15). White further suggests that design researchers focusing on sustainability transitions often overlook the question of politics – especially the kind of politics that center the social demands and needs of workers. Here, bringing politics back into transition design is to recognize that any societal and environmental transition will not be above political conflicts and struggles. This is an important insight. Because of the fact that climate change is affecting every human being in some way or another, there is a tendency to portray transitions through a rhetoric of universalism – one that emphasizes a commonly-shared interest in uniting all humans against climate change. But, climate challenge's harms are not or will not be distributed evenly. Likewise, there is no reason to expect that 'benefits' of sustainability transitions will be distributed equitably unless there is a set of global, regional, local collective efforts to ensure it.

What is the role and place of politics in design as inquiry? If design includes an awareness and engagement with tensions between 'what could be' and 'what ought to be', it has to presume that there are multiple interests, voices, and visions – some of which are incommensurable with each other. Drawing on Chantal Mouffe's political theory of radical democracy, Carl DiSalvo (2015) argues that design should incorporate agonism to make space for such differences:

> From an agonistic perspective, democracy is a situation in which the facts, beliefs, and practices of a society are forever examined and challenged. For democracy to flourish, spaces of confrontation must exist, and contestation must occur. Perhaps the most basic purpose of adversarial design is to make these spaces of confrontation and provide resources and opportunities for others to participate in contestation. (p. 5)

Agonism can help 'politicize' the taken-for-granted ways of thinking about the designed or what is presented as conflict-free, neutral, and universally-beneficial processes of transitions. Design as inquiry approaches social, political, or cultural differences not as obstacles that need to be overcome to reach a solution for everyone. Design with an agonistic attitude will be open to contest others and be contested by others. This means that design as inquiry should inhabit a kind of pluralism without an imposed teleology. It should be ready to acknowledge that fundamental differences might continue to exist between individuals and groups. Thus, agonism can help advocate for spaces supporting alternative design visions that can unsettle status quo in societal transitions.

Agonistic attitude of design is especially helpful in tackling what design scholars call 'wicked problems'. Originally borrowed from the planning literature, wicked problems refer to challenging situations that emerge out of complex social mechanisms. The idea of 'wicked problem' was introduced by Horst Rittel and Melvin Webber (1973) in order to draw attention to the complexities and challenges of addressing planning and social policy problems. Unlike tame problems posed by mathematics or chess, wicked problems lack clarity in its social and technical composition. In a wicked problem, there are many interweaving social, environmental, and conceptual relations. It is not possible to identify a distinct internal logic of a problem. Focusing on one set of relations versus another is not helpful. Furthermore, every attempt to tackle a wicked problem is different from the previous one, as the conditions of the problem and the internal composition of the proposed solution differ in each specific manifestation of the problem.[1]

1 According to Rittel and Webber (1973), wicked problems have 10 major characteristics:
- There is no definitive formulation of a wicked problem.
- Wicked problems have no stopping rule.
- Solutions to wicked problems are not true-or-false, but good-or-bad.

The idea of wicked problems highlights the relational nature of any epistemological project that aims to attend to the situatedness of social issues. For the perspective of a wicked problem, the problem and the solution are in a relationship. One can only think of both, but not one without the other. As such, the framing of a wicked problem makes a set of possible design solutions intelligible and desirable while another set of possible design solutions have been foreclosed. By embracing agonism as a principle, design as inquiry can recognize and value different interests, voices, and visions in framing of a wicked problem.

To be sure, design's agonistic attitude is not simply a proposal of inclusion. It does not aim to create a space in which differences have to fit in predefined categories. Its goal is to transform the space of design in each encounter between different visions. This means being ready to listen to and learn from different philosophies of design. Indeed, some non-western philosophies, including southern African concept of Ubuntu (Kapuire et al., 2015; Chmela-Jones, 2015; Winschiers-Theophilus, 2012) and Daoist concept of Wu Wei (Slingerland, 2000; Loy, 1985), are better equipped to shape a design inquiry in response to the crisis in ethical imagination than most western design frameworks. Captured in the Xhosa expression 'umuntu ngumntu ngabanye abantu' (translated as "I am because we are and since we are, therefore I am"), Ubuntu describes an identity that is intimately tied to the others. Unlike most Western philosophies, Ubuntu privileges neither individual nor society, and instead presents 'related-ness' as the foundation of its worldview. An Ubuntu-inspired design inquiry can provide an ethical grounding based on connections to the community the designed is supposed to serve. Or, take Wu Wei (translated as "non-doing", or "doing nothing"). Characterized as 'action without action', Wu Wei reflects the virtue of finding harmony between oneself and one's surrounding. A Wu Wei-inspired design inquiry can oppose to top-down solutionism of most western design projects, helping engineers and technologists consider non-design and minimal design (e.g., repurposing of already existing material) as valuable strategies. In this regard, Ubuntu and Wu Wei could align well with participatory and co-design methodologies, which are slowly becoming popular in technology design, and direct them towards alternative ways of acting and being with others in the world.

- There is no immediate and no ultimate test of a solution to a wicked problem.
- Every solution to a wicked problem is a 'one-shot operation'; because there is no opportunity to learn by trial-and-error, every attempt counts significantly.
- Wicked problems do not have an enumerable (or an exhaustively describable) set of potential solutions, nor is there a well-described set of permissible operations that may be incorporated into the plan.
- Every wicked problem is essentially unique.
- Every wicked problem can be considered to be a symptom of another problem.
- The existence of discrepancy representing a wicked problem can be explained in numerous ways. The choice of explanation determines the nature of the problem's resolution.
- The planner has no right to be wrong.

Conclusion

The ethics of technology design is integral to many current social challenges. As emerging technologies such as artificial intelligence systems are impacting social life (from economy to education, to health and wellbeing) at a greater scale and pace, more people want to see ethical safeguards to be implemented before a technological system is fully developed. When a technology is out for commercial use, it is often too late to fix the harm that might come from a technology. For this reason, design – not simply consumption – gets more spotlight with respect to the ethics of technology. While nowadays AI ethics is the attention grabber, the questions of ethics have been discussed with regard to the design of manufacturing machines, nuclear reactors, dams, highways, computer networks, and other socio-technical systems. What is new about today's technology ethics debates might be that there is now a larger tendency to see AI and similar emerging technologies through the lens of 'matters of concern' than 'matters of fact'. What is still lacking is design for 'matters of justice'.

Regardless of the type of technology, the fundamental assumption in most ethical design conversations is that the harm that can come out of these technologies could be prevented by designing it differently or not designing at all. This ability to imagine 'designing otherwise' is the very possibility of the ethical in design. I propose the idea of design as inquiry to capture awareness of and engagement with such ethical potentialities in design. Imagining 'otherwise design' has two functions. One is about the future. It generates an awareness of the existence of multiple paths, trajectories, and consequences that are waiting to be actualized. With the opportunity to compare multiple ways in which a design project can unfold, it helps align design with larger concerns and commitments. The other is about the past. It brings attention to how things could have been different if they had been designed differently – like how designing urban spaces otherwise could have generated a different set of experiences for social groups from youth to the elderly. Such counterfactuals are the basis of an ethical imagination informing design inquiries that aim to undo what needs to be undone. Designing for just transitions cannot be without designing away unjust socio-technical systems.

Part III

The four chapters in Part III focus on conceptual and operational dimensions of a decentered approach to science and technology for development. They explore the role that leading with an empathic end-user perspective brings about more appropriate and thus effective scientific and technological interventions.

In Chapter 5, Crystal Brown and Jezabel Aponte examine immigrant processing and integration policies in Panama and Sweden and show what social inclusion means to immigrant groups, which do not often align with the vision of the states. Brown argues that asking the 'right' questions to 'right' people is the key in political science research that aims to support fair and inclusive policies. In her research, as she explains, this means combining statistical analysis with fieldwork.

In Chapter 6, Alexander Smith, Gbetonmasse Somasse, and Robert Krueger bring insights from their work on experimental economics, RCTs in particular, in development projects. They present a diagnosis of why developmental economists, while interested in field experiments, have not created interventions that accurately respond to contextual possibilities and limitations. For them, the missing component of such experiments has been a participatory method with which interventions can be co-developed in a way that aligns with cultural values and community decision-making processes.

In Chapter 7, Ardestani, Burt, and Krueger examine different ways of 'seeing' or measuring poverty and, more importantly, through artificial intelligence tools. By providing examples of their research at the nexus of data science and social science, economic development, multidimensional poverty, and bottom-up multidimensional, Ardestani, Burt, and Krueger offer a decentered framework for technology innovation that prioritizes stakeholder needs by understanding their perspectives in their own words. For them, a bottom-up approach to data collection and design is central to successful implementation of these tools as this is the most effective way to build trust in new technologies.

Chapter 8, offered by Berk Calli and Yunus Telliel, focuses on environmental robotics as a promising area of a decentered technological innovation when it is staged by a public interest technology framing. Conceived of in this way, environmental robotics innovations that extend beyond the simple purview of corporate-led innovation, Calli and Telliel show that robots can help achieve more efficient recycling and recover materials that have critical importance for the society's prosperity and well-being.

https://doi.org/10.1515/9783112218716-006

Crystal H. Brown and Jezabel Aleyda Aponte Figueroa

Chapter 5 Recentering Policymaking: Technique in Policy Design for Migrants and Immigration

Introduction

Across the study of policymaking, some researchers have long accepted that to understand how people respond to policies that are supposed to help them, we need to first understand the perspectives and experiences of those very individuals. The human-centered design (HCD) approach to understanding why people do the things they do has produced a rich body of research. However, many studies about migration often leave out the humans impacted by the policy such as the immigrants themselves and workers helping with the migration process. In this chapter, we argue that a human-centered approach is an underappreciated research tool within migration research. We use HCD to examine migration issues in Sweden and Panama to demonstrate that engaging directly with those impacted by policies can significantly enhance development work. We argue that a human-centered policy approach can provide better insight to complement research designs and development work.

Public policy has long valued quantitative techniques that use statistical modeling to understand human behavior to determine what policy would be 'right' or 'wrong'. In most analyses of migration issues researchers tend to use surveys, engineering designs, assessments of migrants, data mapping of migrant flows, budget allocation figures, etc. to describe the needs of the policy area (Munger and Van Dael, 2020, p. v). Sticking to this methodology can sideline the individuals who are central to the research prioritizing the research questions or the researcher's focus on the policy over the humans themselves. These types of approaches provide information on systemic issues but fail to tell the true story about the challenges, choices, and desires of the people who are impacted by the policies (Munger and Van Dael, 2020, pp. 30–31). The key to HCD is talking with people for whom the design is for, being aware of cultural and social factors, and developing customized research tools. This can be achieved through interviews, focus groups, or vis-à-vis observations in the field or community of interest.

Our point is that research, whether it involves engineering devices, forming policies, or any other study should fundamentally engage with the individuals affected by the outcomes. Ironically, the human-centered approach to immigration policy has thus been absent from debates over major policies affecting immigrant integration and humanitarian workers' needs. The people directly impacted by policies such as

https://doi.org/10.1515/9783112218716-007

immigrants and humanitarian workers are often overlooked in conversations. Instead, a great deal of scholarship has gone into solving migration overflow issues to Western countries and integration factors that help immigrant economically adjust to their new country. While statistical analysis can produce new knowledge and interesting results, it is not the only way to ask questions and fundamentally cannot get at nuanced behaviors.

This chapter demonstrates the importance of how we as researchers ask questions when we seek to understand a phenomenon in each context. Using field work from Sweden and Panama we show that even the most progressive policies on migration fail to address the needs of their constituents. This analysis serves as an example of how engineers, policymakers, and those involved with development work should seek to understand the populations they want to help before implementing development projects or policies. HCD and engineering design for development can address complex problems, which also include migration policies. By focusing on creating solutions to big problems that center human needs, not technological or programmatic needs, solutions can be implemented to meet development goals.

We focus on Sweden and Panama as case studies to understand migration issues from the perspective of refugees and humanitarian workers. Sweden is a developed country in Scandinavia that had a very liberal migration policy beginning in the 1970s. This liberal approach led to the diversification of the populations with more than 20% of the 10 million people in Sweden having Middle Eastern, African, South American, and Asian backgrounds. However, Sweden's legacy of openness came to an end with the large migration flows in 2015 through 2017 due to the refugee crisis in Syria. Many displaced people settled in Sweden, which overwhelmed the welfare state and led Sweden to eventually shut the borders to refugees (Hagelund, 2020). Today, Sweden maintains its closed stance to refugees as it grapples with large poverty issues with its diverse immigrant population. The study for Sweden focused on why there are substantial poverty issues for immigrants from the Middle East and Africa living in Sweden. Using an HCD model helped to expose migration policies that impacted refugees both negatively and positively in Sweden.

Panama is an upper-middle-income country and leads Latin America with the highest per capita income rating as noted by the United Nations Development Programme in 2022. Panama has a population of about 4.4 million people in 2023, divided into five primary ethnic groups: Mestizos (of mixed Amerindian and White ancestry), Indigenous groups such as the Ngabe, Kuna, Embera, and Bugle, those of Black or African descent, Mulattos, and Whites, as reported by the Instituto Nacional de Estadística y Censo. Panama has a large immigrant population originating from North America, Latin America, the Caribbean, and Asia (Bonilla, 2021). As an upper income country, Panama became the second most preferred destination country for migrants in Central America after Costa Rica (Bustinza et al., 2022). Through 2010 and 2020, the Panamanian government implemented 'Crisol de Razas', a pathway for legal temporary residence for migrants living irregularly in Panama (Schlemmer, 2016); by 2020, the number of registered migrants almost tripled, reaching 313,165 making up about 7% of Panama's total population (Bonilla, 2021).

Recently, Panama has taken a more closed stance to legal temporary or permanent residency due to the large inflow of migrations. The number of migrants crossings through the Darién Gap, which is a stretch of land leading from South America to Central America, in Panama is unprecedented every year since 2021. In 2023, the number of migrants crossing doubled reaching more than half a million migrants. The flow of migrants transiting Darien is mixed with economic migrants, asylum-seekers, regional migrants, and extra-regional migrants. As crime has overwhelmed Darien, the region has quickly become one of the most dangerous migration routes in the world. Panama has become both a destination and transit country for people on their way to North America. Most research in this area typically examines migrants in region, but we take a different approach to focus on humanitarian workers' technological needs who are managing the large flow of migrants.

We compare Sweden and Panama due to their significant yet distinct roles in the global migration narrative as both nations have been recipients of substantial migrant inflows during recent migration crises. Sweden has faced challenges in integrating their diverse refugee population which has far-reaching implications for social cohesion. Panama witnessed diverse migratory movements prompting a unique blend of policies that cater to transient and settling populations. The center of the issues in Panama are, of course, migrants, but also humanitarian workers who are on the ground implementing migration policies in times of crisis. We compare these two countries to see how different sociopolitical frameworks and economic contexts influence migrant integration and policy responses. Such a study is important for understanding migration management flows and for strong policy frameworks that can address the complexities of migration and integration.

This chapter contains three sections. The first explains the basic features of human-centered approaches and key challenges to using this approach in migration studies. Our goal in this section is to present the broad ways that scholars have historically used HCD to development design work for policymakers and more specifically in the migration studies. The second and third sections outline Sweden and Panama as case studies. This study shows that human-centered research can help us to successfully develop strategies to solve real-world problems such as poverty reduction and better services for populations in transit.

Human-Centered Design and Public Policymaking Design Approaches

HCD is an integrated process that includes active engagement with those impacted by the development work. This approach requires researchers to place their participants' perceptions and experiences at the forefront of their investigation for analysis (Ryo, 2017, p. 286). In terms of engineering designs, a collaborative approach between development engineering and design studies can lead to more effective solutions in both Western and non-Western contexts. Policymakers in the migration space can shift from problem-solvers to codesigners to broaden their perspectives of design beyond problem-solving. The direction for the study of engineering, design, and development should include a shift toward inclusive practices that prioritizes local communities' knowledge and participation (Telliel and Krueger, 2020). Human-centered engineering design (HCED) focuses on creating solutions tailored to meet human needs and not designing solutions to design something trendy and innovative that might not have an impact. Although not directly linked to migration studies, the principles of HCD and HCED encourage engineers and policymakers to consider the sociocultural context of their constituents which is essential when designing for populations affected by migration. Migration-related projects could benefit from this approach by ensuring that the solutions are not only technically viable but also culturally sensitive and socially inclusive.

Few migration policymakers and policy scholars have used this approach but the ones who have used it in different ways. For instance, Ryo (2017) in her study on why immigrants illegally crossed the border from Mexico into the United States found that immigrants saw themselves as moral people in bad situations due to no fault of their own. The structural issues such as the failure of local or national economies put them in difficult situations. Since they did not want to work for a cartel, they justified their illegal entry into the United States as a moral choice for honorable work that would both take care of their families and contribute to the US economy. They did not see the US immigration policies and enforcement practices as legitimate or fair (in terms of human rights) compared to their moral obligation to their families (Ryo, 2017, p. 290). Ryo's study pointed to the importance of understanding why immigrants choose to migrate illegally and help address the larger issues of push and pull factors that motivate migration. The direct and indirect costs of immigration enforcement are apparent in this study, but we go further to see the implications of policies on immigrants once they settle in a host country and focus on humanitarian workers deployed in transient situations.

Similarly, Min Zhou and researchers (Zhou et al., 2008; Lee and Zhou, 2014, 2015) demonstrated that pathways to mobility for immigrant integration for second generation immigrants were complex by using a human-centered approach. They showed, through in-depth interviews, that the process of social mobility largely relied on structural and cultural demands that affected people and how people responded (Zhou et al., 2008, p. 42). Their research relied on the perceptions and definitions of mobility and success from the perspective of immigrants themselves as the bases of the study. Their research provided why immigrants made their decisions as well as the various mechanisms at play in social mobility outcomes for some immigrants over others. They showed that immigrant parents, particularly Asian and Latino/Hispanic, and their children use ethnicity as a resource to construct a strict framework for success that helps poor and working-class people override their disadvantages (Lee and Zhou, 2014). This emphasis on centering the lived experiences of marginalized groups echoes findings from research on support networks for doctoral students at HBCUs (Brown & Laing, 2025). The study is foundational in that it provides a way to challenge structural issues by talking to immigrants. In the same vein, we wanted to understand how immigrants in Sweden and humanitarian workers in Panama also interpreted their situation considering migration policies that were implemented on their behalf.

In human-centered policymaking there are instances showing its impact though not specifically in migration policy. For instance, the revamp of HealthCare.gov and the Veterans Affairs website following the 2014 implementation of the Affordable Care Act (ACA), commonly known as Obamacare, serves as a prime example of the US government's efforts to better meet user needs. According to Sinai (2020), the overhaul emphasized the necessity of gathering user feedback from both doctors and patients. Researchers involved in the website's redesign discovered that healthcare providers experienced anxiety due to the lack of immediate confirmation when submitting data, with some waiting up to a year for any response (Sinai et al., 2020, p. 6). Moreover, confusion regarding the ACA led some physicians to avoid using the website altogether (Sinai et al., 2020, p. 7). The project's design engineers advocated for an iterative, user-centered design approach, convincing Andrew Slavitt, then Administrator of the Centers for Medicare and Medicaid Services (CMS), of its importance for both the implementation process and the policy itself (Sinai et al., 2020, p. 8).

This strategy led to weekly meetings between CMS leadership and researchers to discuss website improvements and how the goals for Obamacare could be achieved through these digital platforms. Sinai and researchers highlighted that mock websites were tested with doctors, revealing that many were unclear about the purpose of the ACA and how to utilize the website effectively. Collaborating directly with physicians, the team clarified these issues, demonstrating the significant impact of user feedback on policy implementation. The design engineers advocated for the adoption of HCD and agile methodologies in policymaking, suggesting that by prioritizing user feedback and adopting an iterative development process, policymakers can better meet the needs of the public and enhance the overall quality of public services.

Another example of HCD applied to policy occurred in Indonesia. The Asian Development Bank (ADB) was commissioned by the government of Indonesia to study the maritime education and training (MET) system due to concerns about its efficacy. Despite Indonesia's strategic maritime position, MET graduates struggled with certification exams and faced long job searches, highlighting a gap between the supply of semiskilled and the demand for highly skilled maritime workers (Munger and Van Dael, 2020, p. 55). The ADB examined the system from industry, education, and seafarer perspectives through interviews and social listening – a method of analyzing online discussions to gauge public sentiment (Munger and Van Dael, 2020, p. 31). Findings indicated confusion over credible versus deceptive advice in the maritime field, a disconnect further explored through interviews with 58 students and seafarers.

The study also revealed that Indonesian maritime schools often fell short of international standards, with variations in teacher experience and a lack of practical sea training (Munger and Van Dael, 2020, p. 55). Information on entering the field, such as program costs, duration, and certification requirements, was scarce and unreliable, leading some to falsify qualifications (Munger and Van Dael, 2020, p. 65). A critical gap was identified in the lack of available of internships, essential for practical experience, forcing some students into teaching roles due to their inability to secure internships (Munger and Van Dael, 2020, p. 75). The researchers showed that the gap compromised the quality of education, affecting career advancement and perpetuating a cycle of under-preparation among maritime professionals, with many officers stuck at junior levels due to insufficient practical experience and knowledge. It was only through using HCD methods that the researchers gain sufficient knowledge to advise the Indonesian government on potential measures to take.

Through this literature review, we show that prioritizing human needs is important for creating effective policies. The studies by Ryo and Zhou demonstrate the importance of understanding individual perspectives to address the real challenges and motivations of migration and integration. Similarly, examples from the redesign of HealthCare.gov and studies in Indonesia show how engaging directly with end-users leads to more effective policy outcomes. Our findings from Sweden and Panama further illustrate how to use HCD approaches to develop a more nuanced understanding of policies and their impact on people directly in the field. We show the importance of using human-centered research to understand why generational poverty occurs in Sweden as it relates to labor market access, education attainment, and language and cultural knowledge training. And in Panama that the expectations of big international organizations such as the United Nations, World Bank, and Red Cross and small local organizations such as churches have policies that might not be conducive for humanitarian workers on the ground.

Case Study 1: Humanitarian Workers in Panama

Our focus on Panama's Darién Gap also used interviews and observations, which allowed us to explore real-life institutional challenges within the development sector's response to the needs of migrants. Our research looked at the Darién region (Eastern), Panama City (Central), and the David region (Western) which are all key points in the movement of migrants in the country. Panama has become a transit hub for migrants attempting to cross the Darién Gap, a 60-mile expanse of rainforest between Panama and Colombia. This segment is the only unpaved section in the 19,000-mile Pan-American Highway, which extends from the southern tip of Argentina to Alaska. Many of these migrants encounter humanitarian workers during their journey. To gain a closer understanding of the challenges faced by humanitarian workers, we lived in the area for a month doing observations and interviews. We had specific questions in mind about the technological needs of humanitarian workers who are delivering services to migrants. Our research showed how important it is to design technology with real people in mind, especially when dealing with the migration crisis in the Darién Gap.

A decade ago, only a few thousand people traveled through the Darién Gap each year. Today, this once remote jungle has become heavily trafficked. Economic insecurity, political upheaval, rising violence, climate change, and immigration crackdowns have driven half a million people in 2023 alone to pass through the Darién Gap. According to UNICEF, in 2023, around 84% of those crossing the gap were from Venezuela, Haiti, and Ecuador, fleeing economic collapse, political dysfunction, and violent crime. There has also been an increase in migrants from Asia and sub-Saharan Africa aiming to reach the US-Mexico border. A November 2023 report by the Crisis Group estimated that 97 different nationalities crossed through the Darién in the first 7 months of 2023 including significant numbers of Chinese migrants and Afghan refugees.

We selected humanitarian workers to interview for our study using a sampling strategy targeting individuals with direct knowledge or experience related to migration in Panama. We considered seven organizational types to have a representative sample of Panama's development and humanitarian sector: United Nations Agencies, International NGOs, Local NGOs, refugee centers, Networks, and Churches. Our initial context with Fe y Alegria Panama led us to have 18 interviews with humanitarian workers between January 16 and February 10, 2024. The interviews lasted between 60 min and 2 h. Alongside interviews we did two field visits to Darién and Chiriquí Provinces. See map of region in Figure 1. We recorded the interviews and used NVivo for coding and analyzing them. To respect participants' privacy, we anonymized personal and organizational information. The humanitarian workers and their respective organizations are generically referred.

When we assessed the needs of the humanitarian workers, we noticed two types of migrant groups and services provided for them. For migrants who wanted to settle permanently in Panama, aid workers were stationed primarily in Panama City. The humanitarian workers tended to be government workers, local NGOs, and churches helping migrants navigate long-term integration strategies which included immigration legal assistance, educational programs, and career development services. Although the number of people looking to settle permanently in Panama City is not nearly the number of people attempting to go to North America, the city has seen an increase in population by 2% every year since the crisis began. According to Integral Human Development, the population increase in Panama overall is by about 1% in 2024.

Another group of migrants are those who travel through Panama to reach North America. Humanitarian aid workers in this situation at the northern border in Chiriquí and the southern border in Darién operate primarily out of Temporary Migratory

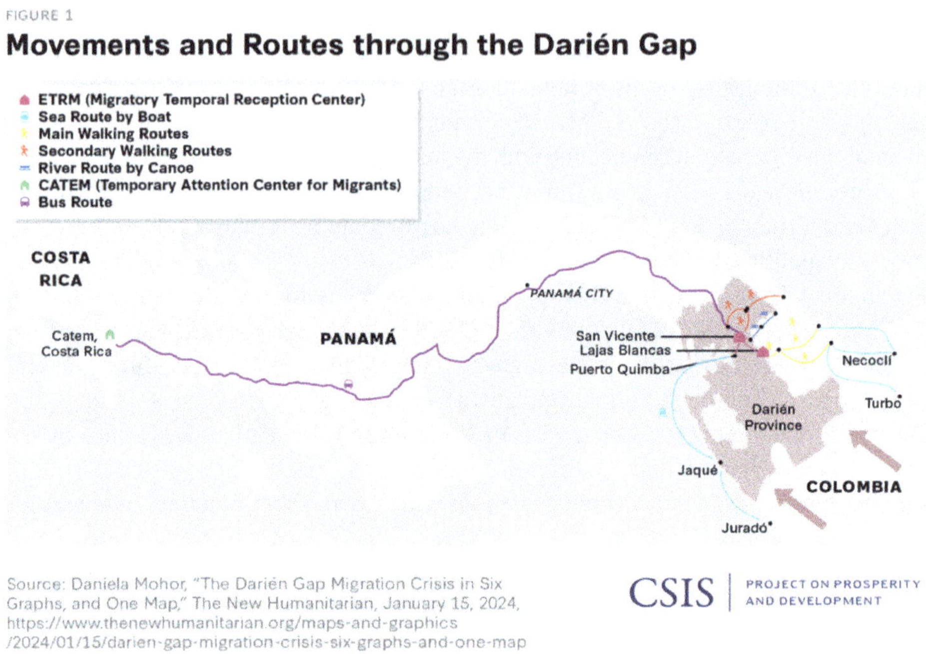

Figure 1: Movements and routes through the Darién Gap.

Reception Stations (ETRMs). Their efforts are focused on providing immediate, lifesaving assistance to the thousands of migrants who pass through these regions. In Darién, humanitarian workers are often the first point of contact for migrants emerging from the jungle. The workers do initial assessments to identify migrants in urgent need of support due to the traumas they faced. At the northern border, the focus is similarly on providing immediate assistance, but there also is a focus on ensuring the migrants get across the border into Costa Rica. Here, humanitarian workers help migrants prepare for the next stages of their journey to North America.

This study focused on humanitarian workers and their needs to assist migrants. Often, development research emphasizes processes or the groups being served such as migrants. However, our focus was on humanitarian workers because we wanted to assess the challenges they experienced in responding to the needs of migrants settled and in transit through Panama. We also evaluated the extent to which these challenges could be addressed with technology-based solutions, specifically information technologies. We present the challenges found in the provision of humanitarian assistance for migrants in Panama and the perceived implications from the interviews and observations we conducted while in Panama. The following sections detail our findings and discuss the implications of our research for humanitarian workers in Panama.

Revealing the Hidden Needs of Humanitarian Workers in Panama

The results revealed that there are two central challenges facing humanitarian workers in Panama that indirectly link to technological challenges. The first is that there is a lack of experienced humanitarian workers in Panama. Several factors contribute to the shortage of experienced humanitarian workers such as low historical demand, retention challenges, inexperienced personnel, threats to safety, and coordination challenges. The lack of sufficient humanitarian workers in Panama has filtered into broader issues in terms of coordination between agencies in Panama, bringing us to the second issue. Coordination problems are prevalent across various organizations in Panama including the United Nations (UN) agencies, international nongovernmental organizations (NGOs), local NGOs, refugee centers, and churches. These challenges were revealed by engaging with humanitarian workers and asking them about their specific needs. Adopting a human-centered approach to the research project revealed information that would have remained undiscovered.

Challenge 1: Lack of Experienced Humanitarian Workers

Understanding the lack of experienced humanitarian workers manifested from our interviews and observations while in Panama. We learned that the shortage of experienced humanitarian workers in Panama is rooted in the country's historical context as a peaceful middle-income country. As an interviewee noted, "The demand for aid workers in Panama has been low in the past three decades. This is because the country, unlike many others in the region, has not experienced any major social, political, economic, or environmental shocks. As a result, a roster of local aid workers did not develop in the country." The low historical demand for humanitarian workers has resulted in very few experienced humanitarian workers in Panama. Hiring is related to potential, not experience, because it is very difficult to recruit and maintain an experienced workforce in challenging environments like Darién. Furthermore, Panama's aid coordination capacities have historically been directed toward supporting other countries in crisis rather than focusing on internal needs. This has limited the development of experienced aid workers within Panama itself.

There is a scarcity of Panamanian workers with the necessary skills in humanitarian development assistance which filters into the ability of workers to provide holistic care for migrants. This challenge was noted by a humanitarian worker who stated, "The personnel who offer support to migrants at ETRMs [Temporary Migration Centers] do not necessarily have the training in psychological care or first aid and do not necessarily know how to proceed when they encounter these cases." Workers might have specialized training in engineering, law, economics, etc. and have good intentions, but they do not have training in terms of working with migrants in crisis. This was exemplified by another humanitarian worker who said, "Managing money or material resources is not the same as managing humanity, and that is a very serious problem for organizations like ours. We [work] with goodwill when what we really need is capacity. If organizations want to have a good team, they cannot settle for goodwill."

Additionally, response organizations in the region cannot offer comprehensive services for migrants because there are very few specialized workers. Often the responsibility for managing the various issues that migrants face is managed by one person who does not have the expertise in all these areas. One worker remarked, "We cannot assume all these roles [human rights lawyer, psychologist, social worker]. We feel that the model we want is a holistic model. We need to have these roles covered so that [we can assist migrants] in discovering their strengths and weaknesses and building a personal plan project." The shortage of experienced workers hampers effective collaboration with other actors and limits the ability to create synergies for integrated service delivery.

Furthermore, humanitarian workers who are mobilized to areas like Darién often face exhaustion and do not renew their working contracts. This leads to a high turnover rate and constant need for new training. One humanitarian worker explained, "Our colleagues generally come for six months, and no one renews contracts, and in fact, in month number four, they are already exhausted. We don't even ask them to renew or extend their contract because we don't see they can do that." Humanitarian workers frequently face severe emotional distress due to being underprepared and overexposed to traumatic situations, risking a deterioration in the quality of assistance provided. As one humanitarian worker noted, "when there is so much rotation, it is a constant challenge because you always must be training people". The long-term impact of these challenges has created coordination and synergy challenges within organizations. The lack of coordination between organizations leads to the second major issue humanitarian workers are facing in Panama, the lack of multisector coordination.

Challenge 2: Multisectoral Coordination

Coordination among aid organizations in Panama is another significant challenge that was highlighted by international organizations. Two main issues were identified: aid duplicity and informal case referral. Aid duplicity is the lack of information sharing among organizations and leads to the repetitive provision of the same type of assistance for migrants. This results in inefficient resource allocation and unmet needs in other areas. As one development worker explained, "Organizations do not always provide information about the programs they run; they do not say the coverage of the programs they have." Another aid worker added, "Regarding the operational challenges, at the multisectoral level, there is a great challenge to avoid duplications." When discussing the issue of issuing vaccinations to migrants, an example was given to illustrate the situation. When a migrant arrives from Darién, development workers provide them with vaccinations and issue a card detailing the vaccines they have received. However, migrants often lose this card. If the migrant then arrives in Panama City and later in David on their way out of Panama, they may receive the same vaccinations again due to the lack of a digitized record. A similar problem occurs with financial assistance, where some migrants receive aid in Darien, Panama City, and David because organizations do not share information with one another creating aid duplication.

Aid duplication is a result of fragmented data management. Different organizations maintain separate databases and do not share information unless they are partners. This is leading to a lack of comprehensive understanding of resources available for migrants, the needs of the migrants, and resources given to migrants. One aid worker pointed out, "There is no catalog of what type of assistance organizations provide, which makes the job very difficult." The lack of coordination also stems from competition between organizations. Organizations often compete for funding and resources, leading to mistrust and reluctance to share information. As noted by a humanitarian worker, "There is much mistrust in the humanitarian world. It is a kind of humanitarian mercenaryism where [organizations] fight for resources and hide information so that their organization can get more resources."

Any type of coordination that occurs between organizations is often informal. Case referral in Panama is often conducted through informal channels without standardized protocols, leading to inefficiencies and potential gaps in service delivery. An example provided by an aid worker illustrates this: "All the organizations in the field know what the referral routes are. If a survivor of sexual violence is identified, according to protocol, we do not inquire but immediately redirect the person to organizations providing healthcare. When that is complete, the organization refers to another organization providing mental health support." All this information is done by text messages and informal channels through humanitarian workers. There is no official system documenting the treatment victims are getting and there is no follow-up plan after initial treatment.

The lack of a formal centralized coordination structure was linked to Panama not declaring millions of migrants moving through its territory since 2015 a crisis. If they had declared this as a crisis, a cluster system would have been used which is the standard for situations like these. The cluster system is a coordination mechanism used in crises to organize the delivery of aid by different humanitarian organizations. The system is structured around different 'clusters', where a specific organization would take the lead and others would follow. For instance, the World Food Programme (WFP) might lead the Food Security Cluster and coordinate all aid related to food for the migrants with other organizations. One aid worker explained, "For the cluster system to be activated, the country has to declare that a crisis exists, and Panama's main problem is that in the case of what is happening in Darién with mobility flows, [the country] has not declared at no time the crisis. So that brings us the challenge of not being able to activate the cluster system."

Discussion: Workforce Development and Technological Integration

Human-centered research in Panama was important in revealing the challenges faced by humanitarian workers. Through direct engagement with aid workers, we learned that while technology is essential for humanitarian workers in Panama, their primary needs revolve around having experienced personnel and better coordination between organizations. The human resources challenge in Panama was surprising given the numerous humanitarian organizations present in the region. However, the primary focus in Panama has been on providing support for other countries rather than addressing their own local needs. The humanitarian workers indicated that developing Panama's local humanitarian workforce would support the needs of migrants and aid workers alike.

Although technology cannot directly address the lack of experienced human resources, online education tools such as Education Technologies (EdTech) can help in capacity building for aid workers. EdTech facilitates distance learning and tailored training programs; it could help humanitarian workers in Panama build their skills and knowledge, thereby improving overall support for migrants. Developing formal training programs that combine theoretical knowledge with practical experience is important. These programs should cover various aspects of humanitarian work, including psychological care, first aid, legal assistance, and social work, to ensure that aid workers are well-prepared to handle different situations. Many organizations already use EdTech to train their workers. Organizations in the Middle East and Africa have used EdTech to develop the humanitarian skills of aid workers. International organizations like the International Organization for Migration and Save the Children are already using EdTech as part of regional capacity-building initiatives. These initiatives could be expanded to include more organizations and tailored to address the specific challenges faced by aid workers in Panama.

Another area for development as indicated by the humanitarian workers in Panama was the need to recruit and retain workers. Our study revealed that developing a well-structured sector of practice that emphasizes professional development and career growth is what humanitarian workers need in Panama. This involves creating clear career pathways, providing continuous training, and establishing standards for professional competencies in the aid sector. Furthermore, attracting a workforce to meet the demand for experienced aid workers can be achieved through incentives. This includes offering competitive salaries, career development opportunities, and recognition of professional competencies through certificates or degrees to attract and retain qualified personnel.

The coordination challenge between international organizations arises from a competitive funding environment and the multiplicity of actors in Panama. This parallels similar gaps in stakeholder engagement found in climate resilience planning in Massachusetts (Brown, McCauley, & Adler-Abramo, 2024). The main implications include aid duplicity, where the same type of assistance is delivered by multiple organizations non-complementarily, and informal case referral mechanisms, which lead to inefficiencies in directing migrants to the best-suited organizations. For the coordination challenge, we suggest that Refugee Assistance Information Systems (RAIS) and Referral Information Management Systems (RIMS) could help address aid duplicity and case referrals. RAIS is a web-based tool developed by UNHCR to store, process, and share information between organizations to track assistance for migrants. RIMS was developed by the Danish Refugee Council to manage case referrals to direct migrants to the best services to support their needs. Using these two technologies would require investment in training and tailoring the systems to meet local requirements in Panama. The issue then would be who would coordinate these efforts in Panama to facilitate the use of RAIS and RIMS. The UNHCR, the Panamanian government, or another international organization?

The human-centered approach we utilized in Panama indicates the necessity for tailored human-focused solutions to support both migrants and the workers assisting them. Technology could help solve these problems, but the bigger issues are about basic human needs centered on capacity building, creating a supportive work environment, and information sharing between organizations attempting to address the humanitarian needs during migration crisis in Panama.

Case Study 2: Immigrant Integration in Sweden

Sweden's integration policies faced criticism in 2015 and later for not fully achieving their goals despite the country's significant diversity increase from refugees since 1945. The problems in Sweden indicated that there were gaps in Sweden's policy goals and outcomes. Integration policies in Sweden supported immigrants maintaining their own cultural identity at the same level as the native population (Soininen, 1999; Wiesbrock, 2011; Borevi, 2013, 2014). Efforts such as funding for native-language education and cultural centers were designed to foster multiculturalism but had some unintended consequences including associations between these centers and terrorism and immigrants creating subcommunities in Sweden (Benmelech and Klor, 2018; Basra and Neumann, 2016; Normark, Ranstorp, and Ahlin, 2017; Ahmed, 2014; El-Tayeb, 2018). These communities where immigrants dwelled were seen as fostering high crime rates, causing long-term social cost, and creating security threats for Sweden (Naper, 2022; Stretmo and Hammarén, 2022; Basra and Neumann, 2016; Short Jr, 2018; Williamson, 2006). The focus of this project was to understand why poverty levels are so high amongst immigrants in Sweden when there are welfare programs to prevent poverty. To get to the heart of this issue, we took a human-centered approach and spoke to the immigrants and observed their communities.

Swedish Case: Methodology

From August 2017 through July 2018, we conducted a mixed-method study on labor market integration, incorporating surveys, in-depth interviews, and participant observations. We actively involved immigrants in every stage from designing survey questions to selecting observation neighborhoods to ensure a human-centered approach in our study. We administered a 15-min survey online in Swedish and English, targeting both immigrants and native-born Swedes to gather diverse perspectives on integration. For observations, we chose to live in and frequently visit immigrant-dense areas in Stockholm and Malmo providing deeper insight into the immigrants' experiences that surveys alone could not capture. All the interviews were conducted with immigrants who had lived in Sweden more than 7 years.

The survey showed that of the 567 survey participants, 42% completed the entire survey. We focused our analysis on these complete responses, excluding two underage participants, resulting in an analytical sample of 236 respondents see table 1. Table 2 presents the demographic characteristics of the sample of respondents; over half (51.69%) of the participants self-identified as females, 47.46% males, and 0.85% as other. This sample mainly consisted of citizens or permanent residents, with immigrants making up 18.22%, nearly mirroring Sweden's demographic breakdown, where 80% are native-born Swedes and 20% are people with immigrant backgrounds.

Table 1: Demographic characteristics of the sample (N = 236).

Variable	Category	Frequency	Percentage
Gender	Female	122	51.69
	Male	112	47.46
	Other	2	0.85
Legal Status	Citizen	221	93.64
	PLR	5	2.12
	RP	5	2.12
	Other	5	2.12
Age Range (Year)	18–34	97	41.10
	35–54	92	38.98
	55+	47	19.92
Status	Immigrant	43	18.22
	Native Born	193	81.78
Total Sample		236	100.00

Note. Table created by the Crystal Brown from data collected for the current study. PLR = permanent legal resident; RP = residence permit; PWMB = person with a migratory background.

Barriers to Integration for Immigrants in Sweden

The first step to analyzing the data was to look at the question that asked participants about the obstacles immigrants encounter that prevent them from fully integrating in Sweden. As outlined in Table 2, the most prominent barriers to integration according to immigrants was finding a job (52.4%), language ability (40.5%), and cultural barriers (26.2%), while racism and discrimination seemed to play less of a role at 16.7%. Understanding Swedish culture ranked number 3, but according to immigrants we interviewed, having a shared culture suggests belonging and a position in the society. Conversely, if someone does not have this "culture" they do not "possess profit and preferment" (Bennett et al., 2009, p. 258). Culture, accessing the labor market, and speaking Swedish are inherently related because, if one cannot speak Swedish, it will be difficult to find a job and understand the culture: most immigrants also said that knowing Swedish helped them find a job, while those who could not speak Swedish were hindered in their ability to work. Our initial analysis pointed to job access, language barriers, and cultural integration as significant challenges to immigrant integration, with employment access being the primary barrier.

Table 2: Integration barriers identified by immigrants and native citizens.

Factor	Immigrants	Immigrants (%)	Native Citizens	Native Citizens (%)	Total	Total (%)
Racism	7	16.70%	52	29.10%	59	26.70%
Language	17	40.50%	107	59.80%	124	56.10%
Raising Children	7	16.70%	53	29.60%	60	27.20%
Job	22	52.40%	91	50.80%	113	51.10%
Housing	7	16.70%	49	27.40%	56	25.30%
Services	3	7.10%	30	16.80%	33	14.90%
Culture	11	26.20%	67	37.40%	78	35.30%

Note: Table created by the Crystal Brown from data collected for the current study. Percentages indicate the share of each group that identified the factor as a key barrier. "Native-born" refers to individuals born in Sweden; "immigrants" refers to those born outside Sweden who have lived in the country for at least seven years.

Given that immigrants ranked finding a job as the top barrier to integration, we examined if a job (and income) truly mattered in Sweden for integration. Employment has become a central issue to integration in Sweden based on an analysis from the OECD in 2007 and 2015, showing that Scandinavian countries are among the worst performers in terms of immigrant participation in the workforce. Consequently, immigrants have lower incomes, higher poverty rates, and are overrepresented among government allowance recipients. Immigrants' success in the labor market depends on many factors related to their self-determination, country of origin, circumstance of migration, class, and the destination country (OECD, 2015). Furthermore, demands for formal education qualifications in Western countries could place immigrants who do not meet these standards at a disadvantage. Finding a viable job has become the key indicator of integration. To this end, Sweden implemented new integration programs to better facilitate labor integration for refugees specifically but no other immigrant groups.

Why Are Poverty Rates High for Immigrants in Sweden?

After seeing the survey results, it was important to know what impacted immigrants in their own words. We use a human-centered approach to explore why poverty rates are high amongst immigrants in Sweden according to their experiences. Our findings suggest three main obstacles: refugees face long waits to work, recognition educational degrees, and lastly social rules and culture for some immigrants can be difficult for immigrants depending on their country of origin. Refugees have a lengthy integration process compared to other immigrant groups, which delays their labor market entry and can lead to long-term poverty. There are two groups of immigrants, labor migrants from non-Western background and refugees, who had higher poverty rates in Sweden. In this section, we will talk about refugees and immigrants from non-Western backgrounds and their experiences with labor market integration. We separate these groups because refugees have an official process of integration that is different from labor migrants with non-Western backgrounds. In some cases, integration was easier for labor migrants and in others it was not.

Integration Issues for People with Refugee Backgrounds Living in Sweden

The group that faced the most barriers to economic integration were refugees. Sweden has a lengthy process for refugees before they can start working. We found that refugees could not join the workforce quickly because the Swedish government required them to learn Swedish and attend basic adult education courses. Swedish language school for foreigners is the language program that all refugees are required to attend and can take 1–4 years depending on the skillsets of the learner. The Swedish government also requires refugees to get at least an adult basic education at Komunal Vuxenutbildning (Komvux), which is equivalent to 3 or 4 years of high school in the United States. These programs aid refugees in becoming integrated in Sweden both in terms of learning the language and developing basic educational skills. After complet-

ing the adult basic education program, refugees can apply to a bachelor's program in Sweden. Unfortunately, Komvux has a negative stereotype associated to it and degrees obtained there are not highly valued. Therefore, people who graduate from Komvux have difficulty securing higher paying jobs.

Additionally, some refugees are educating (or re-educating) themselves for about a decade before they can access the workforce. Many people we interviewed mentioned their positive experiences at Komvux, but they also expressed that the program needed to integrate them into the workforce faster. The language courses and Komvux are required for refugees to receive their monthly stipends to cover their living expenses. However, the stipends are lower than Sweden's average wage, which places the refugee and their descendants at poverty level for a long time because they are often going through the educational training and cultural immersion for a decade. Their wage-earning power is placed at a disadvantage far longer than non-refugees.

University education obtained in Sweden can also be a barrier to labor market integration for refugees. We interviewed an Iraqi refugee, whom we will call Zain, who earned a bachelor's in business from Stockholm University. He moved to Sweden in 2003 as a teenager due to Iraq War. He went through all the language and basic education requirements before entering Stockholm University. Upon graduation, he had a government instegsjobb at a company, but when the government subsidy ended after 24 months, the company did not hire him permanently. They hired another refugee to give them work experience. Instegsjobb gives refugees experience in their field of study; however; companies rarely hire them permanently because of the incentive – to get the 80% subsidy – from the government for each new worker. Zain could not find a job at a big company, so he worked as a gas attendant. He had worked there during college and knew that he could return because he knew the owner. Zain felt that a big company would not hire him because his name identified him as not being ethnically Swedish and the fact that he came to Sweden as a refugee. He argued that the systemic factors embedded in racism hindered his ability to find lucrative work. Even with his previous work experience and his Swedish university degree, he felt that the system would never allow him to get a higher-paying job.

Secondly, we understood that achieving labor market mobility in Sweden often depends on obtaining education abroad in a Western country or getting a degree in a highly desired field (i.e., science, medicine, engineering, etc.) in Sweden. Interestingly, refugees who obtain a degree in a highly desired STEM field or study/work in a Western country such as the United States, Canada, and United Kingdom found gainful employment. Another enlightening interview was with Bisrat (not his real name) who came to Sweden as a refugee with his parents from Eritrea in the late 1980s. His family initially lived in a small town in Scania, a region in Southern Sweden, but they moved to Malmö when he was a preteen. His formative years were in the poorest neighborhoods of Malmö known for high crime rates, unemployment, and diversity. Despite the odds, Bisrat had a high-paying job. He and his wife, also of Eritrean descent, obtained their bachelor's and master's degrees in the United States and returned to Sweden to take management positions. He worked his way up from the poorest neighborhoods in Sweden, which he called 'the gutter' to get where he was in life. We realized that Bisrat's success stemmed from his 15 years of education and work experiences in the United States, making him a highly skilled and desired worker and expert in his field. Bisrat believed that he was successful in Sweden because he understood the Swedish way of life and felt comfortable about his Eritrean background. We will discuss cultural background further later in the chapter.

Like Bisrat, many of the gainfully employed refugee women studied outside of Sweden or got a degree in a lucrative field. A female Iranian professor we interviewed came to Sweden as a refugee in the 1970s. She got her degree at a university in Sweden but was not considered for a professor position until she returned from a 3-year postdoctoral program in Canada. Her work and expertise as a scholar in the Swedish system was not recognized until she validated her skills by working in Canadian academia. Another woman of Eritrean descent moved to Sweden as a refugee when she was 1 year old. She earned a bachelor's degree in computer science from Uppsala University. With computer skills being in high demand and her degree from one of the top universities in Sweden, she quickly got a high-paying position. Her parents, however, never achieved her level of work success. Her father was a taxi driver and her mother was a house cleaner and homemaker – both completed language training and Komvux adult education courses. When we asked her how she achieved so much, she said it was due to seeing her parents working so hard for minimal pay, her computer science background from a top university, and her ability to understand and embody Swedish culture while also maintaining her ethnic identity.

Lastly, we learned that cultural conforming practices such as Jante Law and perceived common descent play an important role in integration. The Law of Jante is a code of conduct that requires people to be modest, not overly ambitious, or do things out of the ordinary – anyone who does not fit the model finds it difficult to belong. We observed that institutions in Sweden have this notion of conformity built into in them. Jante Law promotes the notion of having just enough and not thinking that you are special or better than anyone else around you. We saw that Jante Law also ties into the success of social welfare in Nordic countries: because people are used to living modest lifestyles, they are more willing to share in high taxes to benefit the good of the society. Swedish society holds anyone who transgresses against Jante Law with suspicion and hostility, as it goes against social stability and the sense of harmony. Immigrants who can understand these concepts of social cohesion do very well in the society. Immigrants who do not adhere to these rules are outsiders. A barrier to integration for immigrants is not understanding required behaviors and expectations in Sweden; however, common descent is just as important.

Dr. René León Rosales, who is the research director at the Mångkulturellt centrum, an influential cultural center in Stockholm, suggests that issues of integration reflected issues of skin color. Essentially, people who were born in Sweden or born abroad but are now citizens, whose skin color do not have the normative Swedish physic (i.e., light skin, light eyes, and light hair) are assumed not to be from Sweden. He said, "Often many educated people [and high-income earners] feel really frustrated by how they are treated. Although they are well-educated and although they're really smart, they find themselves in [uncomfortable] situations where people assume." These uncomfortable situations happen because they do not fit into the specific mold of Swedish culture or share common descent with Swedes. For Dr. León Rosales the everyday world is managed by patterns where bodies are connected to places and landscapes, who fits is dictated by how well their physical bodies fit in with the landscape, and in the case of Sweden, White bodies tend to fit and other bodies do not. As Dr. Nira Yuval-Davis outlines in her book, what is required from a specific person to belong in Sweden is common descent (or the myth of common descent), common culture, religion, or language (Yuval-Davis, 2006, p. 209). The everyday experiences of people with darker skin suggest that they are never fully integrated into Sweden. Their race and ethnic background do matter in everyday experiences, especially as it pertains to feeling included in the workplace.

Discussion

Understanding the experiences, beliefs, and perceptions of immigrants, more specifically refugees, may help us recognize where immigration policies have gone wrong. Using a human-centered approach reveals three reasons why refugees remain economically disadvantaged. Firstly, refugees experience everyday racism (macroaggressions) in which employers give preference to ethnic Swedes, people with genotypes that are like Swedes, and immigrants who are educated abroad or with highly desired skills. Secondly, obtaining high-paying jobs is difficult in Sweden because refugees spend a significant time learning Swedish and attending basic adult education programs, placing them a decade behind their peers. In addition, those who earn a degree in a non-STEM subject face strict competition with Swedes. Employers in overly saturated fields such as business administration are more likely to hire a person who is culturally and racially similar to Swedes in a competitive situation of employment over a refugee perceived to be vastly different. We would not have learned of these factors without speaking to refugees about their experiences.

Conclusion

Human-centered research beyond the study of immigration can develop ethical practices in engineering that reflect the needs of those using the product according to their expressed needs. In the case of engineering design for development, human-centered research can help engineers adopt ethical approaches to develop effective designs with input from humans who will be impacted by the design. As mentioned previously, human-centered research is an integrated process that includes active engagement with people who will be impacted by the research throughout the various stages of the project. Although engineering is different from policy research, it similarly uses the basic scientific method to discover new phenomena and is an applied application of knowledge. Engineers are developing, designing, and constructing operating devices, machines, etc. for societal use. New products, devices, and systems will impact humans and thus should be included in the research process, especially in development work in the poorer countries of the world where marginalized people, like refugees, will be impacted. Thus, human-centered research, design, and implementation are vital to the success of all development works.

With every research project, there are limitations. Amongst the various limitations of this study, the fieldwork provided insight into the topic of immigrant integration but did not explain the specific behavior of all immigrants in the world, let alone all refugees in Sweden. Furthermore, gainful employment is not always a perfect indication of integration because the native populations could also suffer from labor market isolation as well. Focusing on socioeconomic integration could create blind spots in the analyses because immigrants could be integrated in different ways such as culturally or linguistically. By no means does this study answer all questions about immigrant integration, but it does provide insight into a complex issue. The overall goal of human-centered research is to provide a positive, long-term impact for the community the study, product, or design impacts. Bringing human-centered approaches to engineering is crucial for the long-term success of any project.

Gbetonmasse Somasse, Alexander Smith, and Robert Krueger

Chapter 6
Creating and Evaluating Co-designed Development Policy

Introduction

Previous chapters have argued that development engineering is not a singular discipline, it cannot be accomplished 'abroad', it must draw from and extend transdisciplinary design approaches, and it must promote integrative thinking and egalitarian processes. In short, rethinking science and technology for development is a practice that integrates different knowledges aimed at producing sustainable outcomes. To do so, it must extend beyond problem-solving in the field to decision tools, innovative policies, and their evaluation. These practices, like that of problem solving itself, is part of the development process, and as such, must incorporate the same egalitarian ethos.

Following the previous chapters in this section, here we bear witness to the history of creating and evaluating policies for development. We place our focus on randomized controlled trials (RCTs) used in experimental economists. We examine the history and recent innovations in this subdiscipline of economics. Then, we turn policy evaluation, in the experimental tradition, on its head. In particular, we make an argument that going to the field is certainly an improvement over desk-based evaluation practices. In our view, field approaches do not go far enough.

In order to understand an appropriate policy intervention or effectively evaluate it, a team must take a collaborative approach, that leverages cultural currency, in terms of setting study boundaries. This means partnering with the expected beneficiaries of the policy intervention to determine successful and sustainable outcomes. Thus, the policy evaluation process is part of a rethought development engineering. To evaluate the intervention a team must take stock of how the intervention was framed and what, from a situated cultural perspective, a successful outcome might be. For example, measuring success in terms of GDP may be futile if, in the process of creating wealth, an intervention destroys a group's capacity to sustain themselves. Take, for example, the work of Edna Wangui who examines the Maasai, a traditional nomadic group of herders who moved with the season following food, water, shelter, and other conditions. However, in the process of making the Maasai 'wealthier' by improving the value of their livestock in externally oriented markets, policymakers promoted livestock substitution, and the new livestock did not complement the Maasai's nomadic lifestyle leading to a number of social problems, not least of which was undermining their capacity for dealing with a changing climate (Wangui 2021).

https://doi.org/10.1515/9783112218716-008

The next section of this chapter examines the role of economics in development history and policy. We then turn to experimental economics as our focus. Here we examine the history of the subdiscipline and explore some of its most innovative practices. After a brief discussion, we take up the main argument of this chapter which is that policy development and evaluation are part of the broader development engineering process and should therefore take a collaborative approach that leverages the cultural currency of the beneficiary.

On Development Theory, Post-colonialism, and the Development Project

The theme of this chapter is collaborative policy evaluation. For generations, economists and political scientists have dominated thinking on policymaking and evaluation. In an earlier chapter by Brown, we heard about the history of this and how some political scientists are beginning to think beyond universalized 'rational' actors coupled with the power of quantitative methods to design policies that actually meet the needs of those they are intended to support. On the other end of the policy process, evaluation, which has been largely the domain of neoclassical economists, the tradition is to employ the same assumptions as policy developers: to understand the intervention as something that is universally legible and can be assumed a priori by policy evaluation experts.

This position has, for a long time, been the target of development theorists and postcolonial theorists. For postcolonial theorists, this notion that policymakers of any stripe can understand from their offices in Washington, DC, or Brussels, the issues that plague countries in the global south, is another arrogant expression of the dominant and universalizing 'modernization' discourses of the global north. Similarly, development studies have critiqued these 'modernist' approaches because they do not actually acknowledge, let alone engage with, the lived experiences resultant from a postcolonial condition. Ed Carr brings together these critiques well in the arguments of his 2012 book *Delivering Development*. His arguments can be summarized as:

- The problems of poverty in the global south are a result of the lack of development.
- Linking the global south with global markets is always a good thing. And, more to our point, linking western resources: medicine, textbooks, institutional foundations, and the like, are 'good' for those in the global south.
- Globalization is a one-way process that bequeaths benefits from northern industrial economies on the under-developed economies of the global south (Carr 2012, pp. 14–18).

Carr develops these arguments into a cogent critique of development as we know it and post-coloniality. His point is simple: the development policy from the global north and links to global markets neither create positive or negative outcomes for those in the global south. People in these places are not as bound up in the Global Economy as we would like to think. In Ghana, for example, globalization may bring batteries from China, recycled clothing from the United States, but their kerosene is refined locally, their food comes from a household farm, and the vessels they use to cook with come from local artisans. Furthermore, global investments in development create opportunities and challenges for local people. Carr notes that we may build a new road to a rural village, but the road itself does not bring higher incomes or greater connection to global markets. The responses to these opportunities and challenges are governed by local decisions based on their own needs and conditions.

This idea is not new. Ignored on this issue by two generations of scholars, the economist Albert Hirschman developed his theory of "The Hiding Hand" (Hirschman 1964). For Hirschman the Hiding Hand did its work when top-down development projects from the World Bank met the social, cultural, and material conditions of a place. Whether it was grazing policy in Uruguay, a railroad in Nigeria, or a pulp and paper mill in East Pakistan, doomed projects brought from the global north were salvaged by local ingenuity and engagement. These local interventions were based on community traditions and ways of knowing and fused with the modern goals of the World Bank Projects. The Hiding Hand represents the nexus of rational, universalized, agendas, and approaches with local customs and know-how. It is these aspects of tradition, local expertise, and culture that policy and policy evaluation fail to comprehend. Before delving into this idea further, we discuss some relevant history of development economics, experiment economics, and RCTs.

A Brief History of the Economics of Development

As was covered earlier in this chapter, international development as a concept can be traced to recovery efforts in the aftermath of World War II (WWII), but the field has its origins long before that in imperialism and colonialism. In its postwar incarnation, development served as a way to bring the underdevelopment of the colonized and the development of the colonizers in the same family with the view that over time the gap between the two sides will disappear. International development is based partly on the premise that different regions, such as countries, have different levels of development. International development is therefore partly about how countries relate to each other.

The development picture around the globe is closely linked to colonialism, which was a key factor in determining early international relations and patterns of international trade, all of which were important for creating the global landscape that existed in the aftermath of WWII.

As an example of the general approach, the Kariba Dam project of the 1950s in the present-day Zambia and Zimbabwe demonstrated what often happened in practice. The large dam on the Zambezi River was built to provide hydroelectric power to surrounding areas. Prime Minister Godfrey Huggins rationalized the project:

> It is vital that we have this cheap power so that we can industrialize and employ our rapidly increasing African population . . . The available land is limited but the African population is not. A permanent solution can only be found through industrialization.

The goal of the project was to stabilize the local indigenous population and promoting migration from Britain to Central Africa (Craggs et al., 2019). Neither happened. However, during construction, the project required 40% of the colonial state's gross national product. Creating the dam also flooded 57,000 local Gwembe Tonga homes (Tischler, 2013).

Another notable example is the so-called play pump, a technology designed to bring drinking water to African communities using the power of children at play. A water pump is powered by a merry-go-round-type device. As children play, the spinning force pumps clean underground water from a borehole into a raised storage tank (Thompson, 2006). With a simple tap arrangement, water can be used by neighboring households to fulfill the needs. The exterior of the storage tank is imagined to be used for advertising, potentially generating revenue for maintenance and repay for some of the installation costs over time. In theory, the idea is brilliant: women can save their energy when compared to hand pumps, children can have fun pumping water into the tank, communities can get freshwater, and the project can be self-sustaining.

The first play pumps were built in the KwaZulu Natal in 1994 on school playgrounds. After Roundabout Outdoor, the company that manufactures the play pumps won the World Bank Development Marketplace Award in 2000, and the play pump gained more attention in the news and was increasingly promoted as an appropriate technology for water supply in sub-Saharan Africa. With support from development agencies and philanthropic foundations, more than a thousand play pumps were installed with the majority in South Africa and the rest in Mozambique, Swaziland, and Zambia and planned expansion to Ethiopia, Kenya, Lesotho, Malawi, Tanzania, and Uganda (UNICEF, 2007).

The rapid expansion of the idea to other African countries was based on the assumption that all young children enjoy playing the merry-go-round game. However, many organizations like UNICEF found the project to be too expensive, difficult to maintain or repair in poor or rural communities, too dependent on child labor, and less effective than hand pumps. For example, many older women, who were not part of the decision-making process that installed the play pumps, found themselves operating the system at their own risks. But was this 'successful' development?

Development Economics and Impact Evaluation

Experts in fields such as economics politics sociology and engineering all have different analyses on the causes of and solutions to underdevelopment in the world. We focus here on the economics of development. Even though neoclassical economics were the first to care about development as the increase in well-being and prosperity, development economics is a more recent branch of economics. Consistent with the initial idea to find and promote recipes that work in many countries, development economics is concerned with a theory that would replicate in developing countries the process that more advanced countries went through. As such, development economics seeks to understand the conditions and policies to elevate poor countries to prosperity.

The search of easy and replicable solutions led economists to come up with various theories of development. Rostow's famous five stages of economic development are a classical example. Rostow imagined development as a linear process that goes from the traditional society to "the age of high mass consumption" with the intermediate stages of "preconditions for take-off", the "take-off into self-sustaining growth", and "the road to maturity" (Rostow, 1954). For Arthur Lewis who won the Nobel Prize in Economics in 1979, developing economies have a dualistic nature. In Lewis's view, less-developed countries have a large subsistence sector with surplus labor and a small capitalist sector. In his model, economic growth occurs when the savings rate increases through an expansion of the capitalist sector which absorbs the surplus labor from the agricultural sector. While Lewis's model provides some distinction between less-developed countries and more advanced ones, the simplification of the development process tends to ignore the complexity that various cultural, social, and economic frameworks bring to the development process. In 1944, Lewis is said to declare the object of a program of rapid economic development to be to 'narrow the gap' in per capita income between rich and poor countries (Arndt, 1981).

Development is more complex, however, and has been split between macro-development economists who focus on economic growth, international trade, and fiscal/macro policies and micro-development economists who study microfinance, education, health, and other social programs (Rodrick, 2008). At the macroeconomic level, development practitioners care about economic growth, environmental sustainability, and a reduction in poverty, inequality, and vulnerability. At the microeconomic level, the key levers are in education, health, finance, technology, and social networks.

Monitoring and Evaluation

As development expands from growth to well-being, its measurement becomes more difficult. The goals of these development policies are not limited to economic growth but include the alteration of incentives of economic agents. To measure progress and assess the effects of these policies, policymakers often rely on monitoring and evaluation. However, monitoring is sometimes confused with evaluation.

Monitoring usually helps policymakers know whether a program is being implemented as was planned. It provides continuous feedback on the program implementation and helps identify challenges as they occur (Baker, 2000). Outcome monitoring consists of measuring an indicator after a policy is implemented. For example, how many students have received a shot after a vaccination campaign. Even when it is well done, monitoring in its simplest form provides no information about the impact of the policy or program if there is no reference to compare to. Outcome monitoring does not tell us about effectiveness. Having a single data points after a policy is implemented does inform on whether the figure is low or high, whether it means progress or regression. Initial improvements in this area using a 'before versus after' approaches lead to a more careful outcome monitoring. This form of outcome monitoring though popular is different from impact evaluation. Identifying the impact of socioeconomic policies matters for the design and implementation of sustainable development policies.

An evaluation is different and can apply to various objects. Baker (2000) distinguishes process evaluation, cost-benefit evaluation, and impact evaluation. We focus on impact evaluation which assesses the final results of various interventions such as programs, projects, policy measures, and reforms on the welfare of communities, households, and individuals (Leeuw and Vaessen, 2009). More specifically, impact evaluation determines whether the intervention has the desired effects and whether those effects are attributable to the intervention, but it can also explore unintended consequences on beneficiaries (Baker, 2000). The purpose is to uncover the causal effects of the policy interventions on key outcomes.

One of the key ingredients for reliable impact evaluation is the availability of a counterfactual comparison which allows the researcher to answer the question 'What would happen without the program?' In the absence of a counterfactual, control or comparison groups are used as a proxy for the state of beneficiaries in the absence of the intervention. The impact of the intervention can then be identified using an experimental or quasi-experimental design (Wong, 2012). Impact evaluation also involves the collection of baseline data before the project begins to avoid data collection bias. This is important because post-treatment control comparisons are only valid if the treatment and control groups were identical at the start of the intervention. However, experimental designs or RCTs are generally considered the most robust methodology of the evaluation by providing a reliable counterfactual.

The Participation Principle

In response to these and other miscalculations, economic development planners have begun to slowly recognize the potential benefits of partnering with local populations in the conceptualization, planning, and execution of development projects. Engineers without borders (EWB) takes this approach. Specifically, they promote that their approach to development is based on 'more than blueprints and measurements'. It is based on real relationships, created through 5-year partnerships. When it comes to latrines, for example, they do not just build latrines. EWB equips communities to build and maintain the latrines themselves (web). Similarly, the Global Sustainable Aid Project (GSAP) builds their microflush with communities and augment them based on the communities' needs and local resources. Finally, it develops microenterprise schemes based on the concept of susu that resonates with communities across sub-Saharan Africa. EWB and GSAP's approach showcases the principle of participation, which is about making sure that the target population of development projects is involved in the planning and execution. This is thought to be critical to long-term success as it empowers the target population to influence and manage their own development. This reduces any culture of dependency (World Bank, 2007).

Imposing development best practices or reforms on a country may produce temporary changes but they can undermine people's incentives and confidence in using their creativity and reinforce their feelings of impotence (Stiglitz, 2002). Recognition of the importance of participation is evidence of a new attitude and ultimately, approach, to international development. In his book 'Whose Reality Counts?', Robert Chambers (1997) stresses the importance of starting with people's knowledge as the basis for planning and change. Similarly, Rodrick (2008) argues that there is an emerging convergence between macro-development economists and micro-development economists around how development policy is done and the recognition of the contextual nature of policy solutions. Mazzurco and Jesiek (2017) provide a list of key principles:

1. Harnessing local resources
2. Collaborating with local champions
3. Creating a competent team
4. Integrating ethics and social justice
5. Building trusting relationships

In a way, participation helps accomplish all five of Mazzurco and Jesiek's principles, but the longer list of five principles provides more insight into how to successfully engage in international development. For new methodologies to be successful in the long term, they must be sustainable, and this means everything must be accomplished using local resources. Partnering with local champions makes it much easier to complete projects using local resources. As a result, the most competent team will be the one that makes local people an integral part of the team from the inception of the project to its evaluation. This is also an effective approach to achieving social justice, and local involvement will, of course, foster the development of trusting relationships.

In environments that embody the five principles identified by Mazzurco and Jesiek, solutions are achieved through a process of co-design, a genuine joint effort of creating solutions in which both outsiders and locals are integral in the creation, implementation, and evaluation of solutions. This approach contrasts the post-WWII approach of outsiders bringing their foreign solutions or best practices from abroad trying to implement effective changes.

While attitudes toward international development have come a long way in the past 70 years, evidence that the new ideologies and principles are more successful than the inappropriate agendas of yesteryear is sparse. One explanation for this could be that the conversation between post-colonialism and decoloniality rarely engages with development theory. As McEwan (2019) explains it, development theory has veered toward the actual practice and implementation of development projects, where decoloniality and post-colonialism have remained in the conceptual world of overcoming binaries such as Occident/Orient, human/nature, master/other. This conversation must take place so that it both reshapes our understanding of the deeply embedded assumptions and of what it means to be 'modern' but it also has implications for the tools that we use to assess development projects.

The purpose of the remainder of this chapter is to do just this. In the next section, we offer an overview of experimental economics with a focus on RCTs. Through this analysis we offer a reading of this subdiscipline of economics that has been widely deployed to assess development projects. We examine the state-of-the-art here through the work of nobel laureate Ether Duflo and colleagues. We then move to an analysis of this work through the lens of post-colonial thinking and then, in a final section lay out a process for moving policy evaluation further in the service of its constituents.

Experimental Economics

Lab Experiments

Many researchers consider Edward Chamberlin's (1948) experiments on the efficiency of markets to be the first economic experiments. Prior to his work, going back to Adam Smith's Wealth of Nations (1776), economists would argue in support of their predictions using theory, and perhaps anecdotal evidence, but could provide little in the way of empirical support.

Chamberlin created markets to test the predictions of demand and supply models. There were buyers with values for the good and sellers with different costs. With the value and cost information, one could derive a theoretical prediction for the equilibrium price and quantity in the market. The test of the theory is seeing what happens when buyers and sellers freely negotiate and trade with each other. Chamberlin found that if buyers and sellers negotiated in pairs or small groups, trading prices and the number of trades often deviated from predictions. However, if Chamberlin announced the trading prices to all market participants, the outcomes better matched the predictions of theory, highlighting the role of information about prices.

Vernon Smith was a student of Chamberlin and followed Chamberlin's work with his own experiments on markets (Smith, 1962, 1964). Smith manipulated the symmetry of buyer values and seller costs and also the way that buyers and sellers negotiated, generally finding that as long as market participants had sufficient information about trading prices, the market outcomes matched theoretical predictions.

In spite of this noteworthy empirical support for Adam Smith's *Invisible Hand* hypothesis (1776) on the efficiency of markets, Vernon Smith is better known for his work on how to conduct experiments (Smith, 1976; 1982). It is for this research that he shared in the Nobel Prize in Economics in 2002 with Daniel Kahneman, who was recognized for incorporating Psychology into Economics (Behavioral Economics). Smith's work on how to conduct experiments was nevertheless quite relevant to his market experiments, which were conducted with fictional goods rather than real commodities. The issue with real commodities is that the researcher does not know how much subjects value the good. Smith proposed inducing values for fictional goods using promises to exchange the fictional goods for real money: a buyer in the market will value a 'widget' at $5 if the experimenter promises to pay the buyer $5 in exchange for the widget. Similarly, a widget costs a seller $5 if that is the amount of money that the seller must pay the experimenter in order to obtain a widget.

The key to inducing values is tying the value of the fictional good to a reward mechanism, such as money, that obeys monotonicity, salience, and payoff dominance. Monotonicity means that experiment participants always want more of the reward mechanism. This is generally true with money. Salience means that the reward mechanism has sufficient appeal that participants will exert effort to obtain more of it. Payoff dominant means that the benefits of the reward mechanism dominate a participant's desire to achieve other objectives, like not appearing to be a selfish capitalist.

Since the groundbreaking works of Edward Chamberlin and Vernon Smith, economic experiments have been used to test a wide range of theories about behavior and to search for fact. For example, when making predictions about how people will make choices under uncertainty, it can be helpful to know their risk preferences. That is, the extent to which they are willing to take on risk, or more often, the lengths to which they are willing to go to avoid or must be compensated for taking, risk.

Holt and Laury (2002) elicit people's risk preferences by having them make a series of choices between two lotteries. For each choice, one lottery has a small variation in possible payoffs ($40 or $32) and the other has a large variation ($77 or $2). When the probability of the high payoff in each lottery is low, one has to really like risk in order to choose the lottery with the higher spread in payoffs. However, as the probability of the high payoff in each lottery goes up, the expected value of the high payoff variation lottery rises faster than the expected payoff of the low payoff variation lottery. Eventually, even someone who is quite risk averse will opt for the high payoff variation lottery. Observing a person's series of choices over the lotteries reveals their risk preferences.

Other experiments have been used to test predictions such as people's ability to play the Nash equilibrium in situations of strategic interaction such as the prisoner's dilemma. In the prisoner's dilemma, each of two players must choose between taking a cooperative action and a self-interested one. The payoffs are structured such that (under the assumption of self-interest) the optimal action for each player is always taking the self-interested action, regardless of what the other player does. The Nash equilibrium of the game is for each player to choose the self-interested action. In decision-making experiments using the prisoner's dilemma, many subjects do not behave in a manner consistent with the Nash equilibrium prediction (see Colin Camerer, Behavioral Game Theory, 2003 for a survey of this literature). While this can be interpreted as evidence against the prediction of Nash equilibrium play, it is also possible that the assumption of self-interest is wrong.

Indeed, many other experiments have examined the prediction of self-interest. In simple bargaining games, Gueth et al. (1982) find that most people are willing to sacrifice some of their payoff to avoid an unequal bargaining outcome. In their trust game experiment, Berg et al. (1995) report that most people are willing to send money to a partner in spite of no guarantee that the partner will send anything back. The amounts sent earn a rate of return and then the partner can choose how much of the amount that they have received to return to the sender to repay the 'trust'. Most partners return significant amounts of money, even though it is not in their financial interest to do so.

In Isaac et al.'s (1984) public good game experiments, group of subjects are all individually given money by the experimenters. They each choose how much to keep for themselves and how much to contribute to a common account shared by the group. Contributions to the common account earn a return and are then divided evenly among all members of the group. The self-interested action is keeping all your money for yourself. In the experiments, most people contribute significant amounts of money to the shared account. This ultimately achieves higher average payoffs for the members of the group.

Experiments have also been used to test predictions in environments ranging from asset markets to auctions. The fundamental value of a financial asset is the stream of dividends that it pays plus any redemption value at the end of its life. Experimental asset markets have been used to replicate the real market phenomenon of asset price bubbles. Traders in the experiments initially trade the asset for less than its fundamental value, but at some point start speculating on capital gains and drive the asset price up above the fundamental value. Ultimately, the asset price 'bubble' bursts and the price crashes back down to its fundamental value (Smith, Suchanek, and Williams, 1988). Using experiments to replicate real-world asset market price bubbles has helped researchers understand the causes and consequences of speculative trading.

In auctions, experiments confirm real-world anecdotal evidence that people often bid higher than they should (Coppinger, Smith, and Titus, 1980), leading to the 'winner's curse': if a number of people bid in an auction, the winner, who has submitted the highest bid, typically pays more than the item is worth. Consumer surplus would generally be higher if the winner submitted a lower bid. On average, real people exceed their Nash equilibrium predicted bids.

Field Experiments

While lab experiments provide control over the data generating process that is helpful for identifying subtle treatment effects, the sterility of the lab means than lab research occurs at the expense of the realism of markets in real economies. This makes the external validity unclear. As such, Harrison and List (2004) recommend field experiments as a compliment to lab experiments.

Harrison and List emphasize that while some experiments may clearly be lab experiments or field experiments, the distinction is not binary, and in fact, there is much room in between. They advocate for classifying experiments using six criteria:

1. The nature of the subject pool
2. The information the subjects bring to the task
3. The nature of the commodity
4. The task
5. The stakes
6. The decision-making environment

Lab experiments are typically conducted on college campuses and as a matter of convenience, students are the easiest subjects to use. While it is typically argued that the decisions of students are at least vaguely representative of those made by the general population, whether this is true in every specific decision environment is unclear. For example, one could imagine that 70-year-olds in the general population might be more patient and less willing to take risk than college students. The extent to which this could interfere with the accurate measurement of treatment effects is hard to determine.

In lab experiments, subjects usually bring no prior information about the task to the experiment. As a result, they must learn everything from scratch. In contrast, in the field, the subjects are often familiar with the task. In a somewhat similar vein, in a lab experiment using a fictional commodity, a subject's value of the commodity must be created within the context of the experiment. In the field, subjects are familiar with the commodity and have some value for it already. The challenge for the researcher is not knowing what this value is!

When it comes to the task, in lab experiments, the tasks are often contrived and require a significant amount of explanation for the subjects to understand what exactly it is that they are supposed to do. This has the potential to lead to confusion, cognitive fatigue, and possibly random behavior when it comes to decision-making. In the field, subjects are typically performing a task that they are familiar with. It is only the incentives or parameters of the task that are being manipulated by the experimenter.

Lab experiments are often over modest stakes such as $10 or $20. In the field, stakes are generally much higher. In fact, they tend to be more appropriate for the task. For example, in a labor market experiment, the stakes are the real wage rates of the participants. This helps to achieve salience and payoff dominance, as discussed by Vernon Smith.

Finally, in lab experiments, the decision-making environment is the lab, which tends to be sterile and contrived. The fact that subjects' decisions are being observed is obvious and this has the potential to affect behavior. In the field, the decision-making environment is realistic. Subjects may be more comfortable in the natural environment and less aware of the monitoring of their actions.

Natural experiments are an extreme example of researchers taking advantage of the 'realism' provided by the field. In such experiments, researchers have no control over the experimental treatment (or manipulation). They are passive observers to a policy change chosen by government or some other authority. A good example of this is that in 1992, the United States Department of Defense began offering significant early retirement incentives to almost 300,000 members of the military. These were real people, who turned out to vary significantly in terms of age, ethnicity, length of time in the military, etc., who were making an important life-changing decision over large amounts of real money. Much information was provided to them about their options.

Warner and Pleeter (2001) recognized the value of the decision data and analyzed it to learn about discount rates, which measure a person's patience for receiving money (a higher discount rate indicates less patience). Those who chose to retire were offered a choice between a lump-sum payment and an annuity. Less patient people would choose the lump sum because receiving the annuity meant your benefits were delayed over time.

Warner and Pleeter's analysis shows that discount rates implied by the retirees' choices are higher than many economists expected, and higher than the discount rates typically assumed in the economics literature on receiving benefits are different points in time.

As the literature on field experiments grew in the early 2000s, Levitt and List (2009) provided an overview of what they argue are three distinct phases in the evolution of experimental economics. First, they argue that Fisher and Neyman's research on agricultural productivity in the 1920s and 1930s is really the first example of field experimentation. Fisher and Neyman conducted RCTs by randomly assigning different plots of land to be in either treated or untreated with an intervention. Of course, there were no human subjects making decisions regarding money, which is likely why these experiments are often overlooked. However, the topic of study was production, which is economics, and the researchers were using the experimental method to identify treatment effects.

Next, in the 1960s, governments conducted social experiments on policy issues such as employment programs, electricity pricing, and housing programs (see Hausman and Wise, 1985 for a review). They would actively manipulate the policies in an effort to identify the effects of policy. In the 1980s, such experimentation tended to revolve more around the effects of policy changes. Academic researchers often overlook these experiments, perhaps because they were not as involved in the work as they would have liked. Some argue that these experiments were in some ways policy creation by 'trial and error' than carefully designed and controlled research agendas, but they were large-scale field experiments.

Finally, Levitt and List argue that the current economics field experiments are really the third phase in a history that spans 100 years. However, the third generation of field experiments indeed have the virtue that the decision environment is typically manipulated in a very careful and controlled way, inspired by the experimental methodologies developed in the lab.

Henrich et al. (2004) conducted bargaining, trust, and public good games in 15 small communities in developing nations to learn about the consistency of self-interest (or deviations from it) across cultures. List and Shogren (1998) learn about behavior in auctions by conducting auctions for trading cards at trading card conventions using subjects who are dealers and buy and sell trading cards all the time. Croson and Shang (2005; 2008) examine how donors to National Public Radio in the United States respond to information about what a previous donor has given, thus learning about peer effects in charitable giving. Finally, Levitt and List (2009) cite the experiments in development economics by Kremer et al. (2004) and Duflo et al. (2006). The former study the effects of a merit scholarship program in Kenya. The latter examine the effects of education programs on HIV prevention, also in Kenya.

Experiments in Development Economics

Development economists have had a different path toward conducting RCTs in the field to that of experimental economists. Many years ago, many development economists used naturally occurring data to measure the effects of policy, but of course, using naturally occurring data to measure treatment effects comes with challenges.

The traditional problem is that if people choose whether they are in the treated or untreated group, one has to expect that those who are more likely to get benefits from the treatment to be more likely to get it. This leads to researchers overestimating the treatment effect.

In some cases, it might be possible for researchers to control the factors that make some people more likely than others to choose to be treated, but sometimes such factors are either immeasurable, or the data is not available. For example, with flu shots, we know that older people are more likely to get the flu shot. They get larger benefits from getting the shot as the consequences of getting the flu are on average more severe for them. We can control for age. When it comes to previous medical history and preexisting health conditions, these are factors that also affect someone's choice to get the flu shot and likely the benefits that they receive from it, but the information about these factors that is available to researchers is normally incomplete at best.

Bertrand et al. (2004) focus on a different challenge when it comes to estimating treatment effects using naturally occurring data. In addition to the problem that people's personal decisions about receiving the treatment can bias estimates of the magnitude of the treatment effect upward, with naturally occurring data, researchers may be prone to overestimating the statistical significance of treatment effects because they overestimate the accuracy of their treatment effect estimates. That is, they believe the standard errors of their treatment effect estimates to be inaccurately low.

The authors explain that the problem stems from serial correlation. If one observes treated and untreated groups over time and uses a difference in differences approach to identify a treatment effect, the researcher's regression error terms (difference between actual and predicted outcomes) are almost always serially correlated. This leads to underestimating the standard errors of the estimate of the treatment effect. Bertrand et al. propose an effective solution (a correction) to obtain accurate estimates, but the issue nevertheless highlights the challenges associated with using naturally occurring data.

Banerjee and Duflo (2009) survey the rapidly growing literature in development economics experiments (specifically RCTs), for part of which Abhijit Banerjee, Esther Duflo, and Michael Kremer were awarded the 2019 Nobel Prize in Economics. Much of this literature focuses on using RCTs to evaluate education policy in Africa. Interestingly enough, the results of RCTs indicate that some policies that intuition might suggest should promote better outcomes actually have very small effects. Other policies that are perhaps not initially as intuitively appealing can in practice be quite effective. For example, many people believe that access to textbooks, which are often shared among multiple students, is a key factor underlying student performance. Glewwe et al. (2009) report that improving access from one textbook for every four students to one for every two students have no significant effect on average test scores. Similarly, many policymakers are heavily focused on student-to-teacher ratios. Banerjee et al.

(2005) present evidence that cutting student-to-teacher ratios in half has little effect on test scores.

On the other hand, simple interventions like deworming treatments can have significant effects on student performance and are highly cost-effective (Miguel and Kremer, 2004). A deworming treatment that cost $0.49 per child per year reduced student absenteeism by 25%. This maps to better test scores than, say, giving teachers compensation for better test scores, creating an incentive for 'teaching to the test'. Part of the beauty of the deworming intervention is that not only does a student who has received the treatment become less likely to be absent, but such a student is also less likely to pass the malady to others, improving the outcomes of students who have not even received the treatment. As such, this simple intervention is highly cost-effective.

Another effective intervention is one that indirectly promotes teacher attendance: teachers are compensated for taking time-stamped pictures of themselves with their students in the morning and afternoon. While the compensation is attached to the pictures, a teacher cannot take pictures with students without being at school. Duflo et al. (2007) found that this compensation scheme reduced teacher absence from 40% to 20%, with significant accompanying gains in student performance.

Reflections

Many lessons can be learned from the literatures on development, development economics and impact evaluation, and experimental economics. First, development efforts have always been well-intended, but have not necessarily occurred for the best reasons. For example, efforts have often been attempts to close the gap in economic prosperity between developed and developing nations or between colonizers and the colonies. Alternatively, development was an attempt to speed the recovery from WWII through growth fed by globalization.

These are not necessarily the best reasons to pursue development. A different outlook would be to try to see every country be as good as it can be. This makes it less of a 'keeping up with the Joneses' race between nations or about a recovery from loss and puts the development agenda in a more positive frame. Addressing the historical 'Why' problem associated with development might also be contrastive for addressing the 'How' problem. Failures such as the Kariba Dam project and the Play Pump initiative are not rare examples and are evidence of a need for policymakers to understand that development must happen differently to be successful.

Indeed, scholars have proposed healthier philosophies regarding how development should occur. The Participation principle emphasizes the role of local stakeholder participation in the planning and execution of development projects. This is good for engagement and ownership over the project and improves how local participants feel about themselves and the project. In the least, it does not create the negative sentiment associated with outsiders coming in and telling you what is best for you (so that you can be more like them). In spite of this insight, however, development scholars to be fast to criticize antiquated, traditional approaches to development, and quick-to-say development should happen differently, but slow to implement the approaches that they propose.

They are not alone though. Development economists have struggled with similar challenges implementing the right policies in the most appropriate way. The traditional approach to development economics has been identifying successful policies in developed nations and taking those policies to growing nations to replicate the same success. Rostow's (1954) five stages of development trace out the path that one would want each nation to follow. Lewis (1944) spells out how moving through Rostow's five stages coincides with capital accumulation and the shift of labor from labor-intensive sectors to capital intensive sectors. The idea is that at early stages of development, the capital stock in low and most labor is employed in labor intensive industries. As a nation accumulates capital, labor shifts toward capital-intensive industries. This is generally what happened in developed nations and there is indeed high correlation between development and the size of a nation's capital stock (in fact, by some definitions of development, the capital stock is a measure of development), but economic development is more complex than these theories would have one believe.

Measuring economic prosperity is an imprecise science and measuring well-being is even more complex. The importance of context and associated sources of nuance are often overlooked by traditional development economists. Further complicating the matter, when it comes to development policy, there is often confusion between what is monitoring and what is evaluation. Monitoring refers to whether a policy initiative is being implemented as planned: Is the intervention occurring as it is supposed to? This is different from evaluation which addresses the effectiveness of the policy: Is the intervention improving the outcomes that we are aiming to improve?

Even with a proper evaluation of effectiveness, it can be very difficult to determine the causal chain underlying why a policy works. The primary reason is that with naturally occurring data, the selection of observational units to treated and untreated groups is normally nonrandom, making the causal effect of the policy unclear. Experimental economics provides a methodology for measuring casual effects using RCTs, but it leaves much to be desired in terms of how policy interventions are created.

The benefits of the methodologies of experimental economics are that they permit a careful and controlled measurement of treatments effects from which one can make causal inferences (due to the random assignment of observational units to treatments). Additionally, there is a good understanding in the discipline of the tradeoffs that come with moving from the carefully controlled environment that is a decision-making lab to conducting field experiments in the real world, the natural decision-making environment, with subjects from the population of interest who perform the task regularly. The subjects bring with them familiarity with the task and they participate in the experiment using real commodities and realistic stakes.

The shortcoming with experimental economics, and experimental development economics by extension, is where the interventions come from. Banerjee and Duflo (2009) provide an honest and very helpful discussion of a variety of issues surrounding the problem of how interventions are created. First, academic researchers often have little sense of what will work well. For example, improving access to textbooks and reducing student to teacher ratios have provided disappointing benefits in terms of student performance. However, Banerjee and Duflo also argue that local NGOs are often 'clueless' too and 'not much better informed'.

Though Duflo and Banerjee argue for creating long-term relationships with local stakeholders, rather than engaging in one-off projects, they suggest that the reason to do so is so that the researcher can engage in an iterative series of experiments in which initially a broad intervention is studied and then later, the researcher uses follow-up experiments to study more targeted interventions and determine why they work. They suggest that creating relationships can give the researchers more influence in convincing the locals what to do. The idea is that the researchers can share knowledge of what has worked elsewhere and advocate for attempting to replicate the results locally.

To be fair, Banerjee and Duflo make a compelling case for building relationships with local stakeholders, the important question is whether this then fulfills the essence of the Participation principle. Additional reason for promoting local participation is to help address common problems or criticisms of experiments. One common issue is the issue of external validity or environmental dependence. Just because an intervention works in one place does not mean it will work with a specific population. The benefit or working with local stakeholders is that they help in the recruitment of the population of interest.

Another issue in experiments is compliance. Is everyone in the experiment doing what you want them to do in terms of using (or not using) the intervention. The thought is that working with locals helps mitigate noncompliance. Finally, there is the Hawthorne effect: some people change their behavior when they know they're being observed. This problem is potentially worse when one is observed by a strange foreigner than by a local person with whom you are already familiar.

Thinking and Doing: The Path Forward

For the past year, the authors of this chapter have been engaged in designing and refining an approach to collaborative RCTs. This section will offer the broad contours of this project to illustrate the points we raise above. For additional information on the project and approach, please see our working paper series on the Institute of Science and Technology for Development's website: https://www.wpi.edu/academics/science-technology-development.

Summarizing scholar Ron Eglash, it is important for African students to realize that they have a heritage about Mathematics, not just singing and dancing, as is popularly portrayed. This notion captures the goal of our work which is orthogonal to fixing problems (i.e., deworming) for why students are not learning math to Western standards in Africa. Indeed, healthy students learn better. However, students also learn better when they understand their agency and connection to the material that they are ingesting. Mathematics isn't something that was 'discovered' by Pythagoras and the ancient Greeks; fractal geometry was not 'discovered' by German Mathematician Georg Cantor (1870s) or Benoît Mandelbrot in the 1970s. As Eglash (1997) has pointed out, fractals have been part of many cultures in Africa for millennia – they just are not getting credit for it. As Oxford historian Hugh Trevor-Roper stated late in the last century, "Perhaps in the future, there will be some African history to teach. But, at present there is none: there is only the history of Europeans in Africa" (Phillips, 2006). For Mavhunga (2017) colonialism from the 1850s to the late twentieth century killed, disrupted, or delegitimized African sites of innovation. Yet, again following Eglash, a more realistic view is the one that reveals a resilient and innovative capacity of indigenous traditions to adapt to new conditions so that they become part of everyday life (Eglash and Foster, 2017). Eglash and Foster suggest that Africans look to themselves for inspiration rather than importing other people's cultural values (Eglash and Foster, 2017, p. 135). It is in this spirit of innovative integration that we have embarked on a project that brings a mathematics learning platform developed in the United States yet, in contrast to the adopted pedagogies and desired outcomes of their former colonial masters, helps teachers and students to engage in mathematics in their own terms.

The learning platform 'ASSISTments' was developed and been improved for the past 20 years by the husband and wife team of Neil and Cristina Heffernan. Beginning their careers as teachers in Baltimore they identified a need to provide more effective feedback and data-driven information that teachers could use to inform their instruction. ASSISTments is a portmanteau that combines tutoring assistance with assessments. Two foci make ASSISTments effective: First, it is a tool, not a curriculum. The tool provides teachers with real-time data about how students are addressing their problem sets. Second, is their mantra "put the teacher in charge, not the computer". Through ASSISTments teachers can examine which concepts individual students or groups of students are struggling with. They take this information to create bespoke learning scenarios for their students. To date, ASSISTments has served 500,000 students and some 20,000 teachers in North America. Any textbook can be placed or uploaded onto the platform, which has downloadable and online versions. To be clear: what makes ASSISTments unique, and therefore useful here, is that it does not teach math but provides teachers insights into student learning.

If we return to the chapters, and indeed, the book's opening argument that development must incorporate different forms of knowledge it is going to produce sustainable outcomes. Taking that point to RCTs and ASSISTments we begin to see an actionable approach that is capable of pointing to how student learning could be improved. Simultaneously, it can be the basis for an exciting new form of RCT because it meets students where they are rather than where some pale colonial teacher's ghost thinks they should be. As a result, students will learn and be rewarded for learning concepts rather than just arriving at the right answer. Finally, the data gained from these studies will be able to inform policymakers in their efforts to design and enact policies that support student learning. So, like Duflo and colleagues, it may not be the textbook, or even shoes, that improves learning outcomes, but rather the actual approach to learning.

Collaborative Approaches for Assessing Learning and Designing Policies

To begin this work, we recruited a small number of teachers in Ghana for a trial study on the feasibility of implementing ASSISTments in Ghana, its usefulness to teachers, and effectiveness for students. We recruited 17 teachers to participate. The teachers represent five regions: Greater Accra, Ashanti Region, Eastern Region, Bono Region, and the Upper East Region. Teachers recruited 38 students to participate in the initial trials, which took place between October 1 and 25, 2020.

The data from our interviews and focus groups, as well as our ongoing work with this cohort of teachers, suggests that ASSISTments can be effective for Ghanaian mathematics teachers and their students. Our next step is to carefully design RCTs to measure the efficacy of ASSISTments.

Limitations put on research travel have limited our ability to carry out these trials. Further limiting our work is home-based schooling in school districts where we have relationships. We cannot access students from different cultural backgrounds, especially Ghanaians, which number over 20,000 in our community because of COVID restrictions. Figure 1 summarizes the process we have outlined to carry out the work once restrictions ease.

These RCTs carried out in different regions of Ghana will employ ASSISTments as the assessment tool for determining mathematical literacy. With the flexibility of the ASSISTments platform we can test different treatments to determine how students improve their mathematical literacy, not just learn to apply their knowledge perform well on classroom examinations or national proficiency tests.

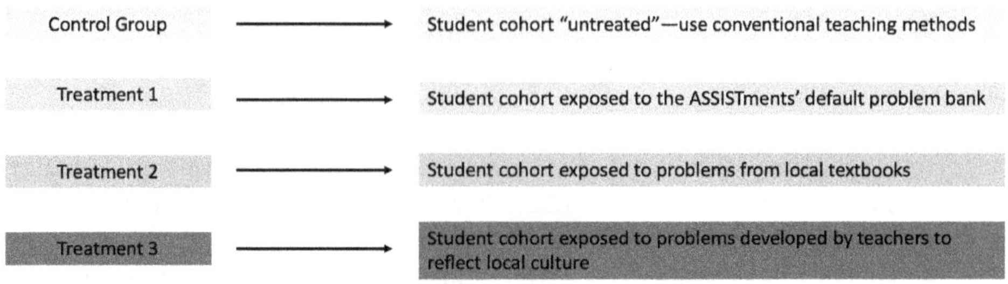

Figure 1: Randomized Controlled Trials Using ASSISTments.

Zahra Zarei Ardestani, Martín Burt, and Robert Krueger

Chapter 7 What Do Artificial Intelligence and Social Science Have to Do with Poverty Alleviation?

> Information is to data what wine is to the vineyard: the delicious extract and distillate.
> –David Weinberger (2011)

> Data have become a key resource in the modern world, mediating how business is conducted, governance enacted, knowledge produced, and the practices of work, consumption, travel, play and domestic tasks.
> –Rob Kitchin (2021)

Introduction

For over a decade big data, neural networks, and more recently, artificial intelligence (AI) have become all the rage in data science, some of the social sciences, and industry from government, healthcare, e-commerce, and even in education and social services. It's not just industrialized countries who will face the impacts of AI and the digitization of society. Indeed, 5 years ago, the United States Agency for International Development (US AID) developed a digital strategy that has evolved, in July 2024, into a policy that will impact various agency initiatives from poverty interventions, governance, and cyber security (US AID, 2024). How will this new generation of science engineer new outcomes? This chapter contributes to the discussion on engineering and development by examining one of the most pernicious problems of development – poverty.

We examine the commitments developers and practitioners make when using big data and AI to 'understand' poverty. Data and their derivatives, indexes, are created by researchers and software developers and used by policy-makers. Many times, these data (1) were originally collected for a different purpose, (2) are outdated, and (3) are suggested to be representative of an entire 'population'. Look, for example, at the advertised data sources currently used for the Oxford Multidimensional Poverty Index (https://ophi.org.uk/). Some go back to 2009 and, collectively, the median year is 2018. Eight-year-old data is one challenge. More significantly, is other intervening variables such as COVID-19, civil wars, drought, and famine. Thus, when such 'data driven' policy is conceived than implemented it may be working to solve a problem that does not exist, no longer exists, or maybe, never existed. Worse still, it might deploy a solution that does not match the problem.

This chapter examines the emergence of AI in one aspect of development: poverty. Our purpose for this is twofold. First, our analysis offers a critique in the sense

https://doi.org/10.1515/9783112218716-009

that Max Horkheimer wrote about in 1939 when he stated: "By criticism, we mean that intellectual, and eventually practical, effort which is not satisfied to accept the prevailing values, ideas, actions, and social conditions unthinkingly and from mere habit" (Horkheimer, 1939, p. 338). Kitchen (2021(a)) calls this 'critical data studies'. Our contribution to this discussion is twofold. First, we examine the philosophical principles, epistemologies, and scientific practices, behind data science, and especially its goal of shaping social things: norms, policy, and practices. We then sharpen our focus on the use of data science and AI in development with a particular focus on poverty. After offering our critiques of data in development to address poverty as currently experienced, we reassert our analysis to offer another, complementary, approach to science, technology, and development by bringing an interpretivist perspective, drawn from social science and incorporated into data science and the use of AI methods tools such as machine learning (ML).

To develop our argument, we will use the 'Poverty Stoplight' developed by researchers at Fundación Paraguaya and coauthor Burt as an exemplar for how we can begin to close the gap between science and technology for development and the actual needs of the purported beneficiary. A penultimate section offers a comparative analysis of the different approaches for redressing poverty. For discussions on cognate topics including good governance, AI, and equity, and the like see Telliel (this volume).

Data, Social Science, and Data Science

Society has been 'datafied' for nearly a generation. Search engines such as Yahoo, Explorer, Alta Vista, and Ask Jeeves understood early on in the internet revolution that data was going to be a new, or even, the new commodity. Innovations in memory and storage, going from cassette tapes, floppy disks, and zip drives to server farms supporting 'the cloud'; created easily accessible storage for the petabytes of data worldwide that flowed freely beyond the internet giants such as Google, Amazon, and Facebook. Computer languages such as Perl, MySQL, and Python have created opportunities for organizing, collecting, parsing, and analyzing these big data. In the social sciences 'R' has brought coding back from the dead as SAS and STATA lost favor to SPSS, which could conduct a factor analysis with a click of a button. R is more than a statistical package as it has analytical tools as well as supporting algorithms for scraping and visualizing data. Data scientists seek to improve, optimize, and problem-solve using these tools and available data.

While social science and data science have the word 'science' in common, they are not siblings but distant cousins. For now, social science seems to be the cousin no one really wants to talk to. These communities do things differently, such as the scientific method, and may be used differently at different times of their practice. Data scientists have many different objects of analysis. For example, a data scientist may use a mobile phone's gyroscope to determine someone's health status, conduct surveys to measure the impact of a policy, and use images captured by mobile phones to make medical diagnoses, they may look at criminal records to predict the recidivism of convicts. Using the same conventional wisdom, behavioral scientists running university labs often perform experiments on college students and generalize their findings to society or a larger social group.

Ontologically speaking, here, data scientists and social scientists are close: there is a single subject, an essential human, an archetype, that can be extrapolated to species or other population. Many know this as a positivist philosophy of science. The method here is the same one many have been taught since they were in primary school. Positivist social science, and much of science, for better or worse, rely on the scientific method as shown in Figure 1.

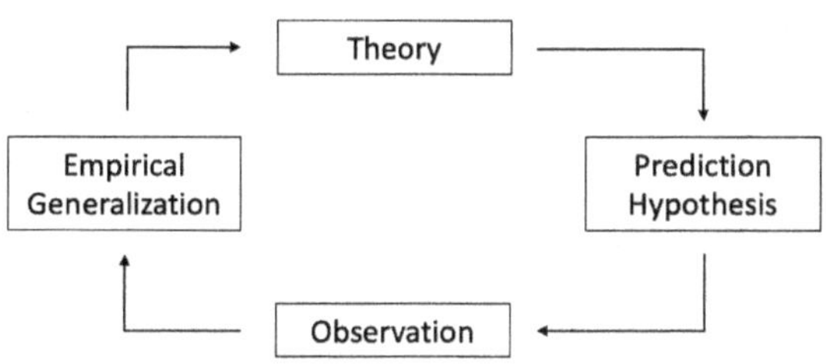

Figure 1: The Scientific Method.

How does data science differ from this? Data science is less interested in explanation and more interested in correlation (Anderson 2008). The idea being that if you have enough data correlation equals explanation. Bear with us while we go off topic for a minute to illustrate the implications of this. Imagine that a data scientist is interested in predicting shark bites off a beach. They could try to count sharks – more sharks, greater risk of bites. They might estimate the number of beach goers on a certain day. More people, more risks of a shark bite, right?

Either of these approaches have some scientific plausibility. Yet, they would require costly equipment and time-intensive data collection. Why do this when you can identify a logical proxy that is available and easily accessible? For a data scientist, a simple proxy for determining the risk of a shark bite could be the amount of ice cream sales. The core assumption is that the more ice cream sales, the more people at the beach, and therefore, more people at the beach can predict the risk of a shark attack. Simple.

Let's return to how different scientists look through different lenses and at different units of measurement. A biologist would argue that the season is important. The biologist could successfully argue that an important intervening variable might be that sharks concentrate near sources of food – seals. Seals congregate during times of the year in areas where they can birth their calves. This has nothing to do with ice cream. Similarly, a chemist might examine the impact of climate change on deep ocean waters. Climate change is known for warming and cooling the top layers of ocean waters that could attract sharks. Furthermore, the sharks might not be present in the traditional locations or seasons. Ice cream does not come into play here, either.

The point here is that a study of ice cream sales and the relative risk of a shark attack is only based on the correlation of observations, and not a causal link. The data scientist is looking for the largest datasets possible to train models to examine specific questions. A data scientist is agnostic about where or why data is collected. They are concerned with whether the data available is suited for their model. The shark attack example approaches science differently. It uses proxy data, easily accessed, that lacks relevance to sharks, their ecosystems, or geophysical changes. In the examples that follow, the questions researchers asked are based on theory that has undergone rigorous testing or significant conceptual clarification.

Some social scientists might agree with this approach while others do not. For those World Bank economists of the 1950s and 1960s who defined success through GDP might agree. In contrast to this conventional wisdom of the time, a few economists, such as Albert Hirschman, noted that economic models and indicators of development do not apply to low-income countries because of different social experiences and his concept of the 'hiding hand' (see the Karnapuli Paper Mill example from the introduction). By analogy, there is no concern with local human behavior (i.e., the local labor force, ethnic tension, and the like), and there is no effort to understand dynamic and entangled relationships within an ecosystem.

Today, social scientists and humanists have raised important questions about data science in terms of ontology, epistemology, methodology, and ideology (Kitchin 2021). Theories driving social sciences are not the same as in data sciences. In our examples above we showed how data that seem irrelevant to a question can be used to identify correlations between phenomena. Correlation is not causality. Let's look at Kitchin's (2021) areas of difference in Table 1.

In this section we have sought to show some of the differences between how data scientists and social scientists ply their trade. Our purpose was not to suggest one is better than the other, but that when studying humans data scientists need to be aware of the commitments they are making; otherwise what looks like a good idea could be a disaster. The history of development is riddled with anecdotes of disaster. More importantly, we created a space where they can be understood as complementary rather than oppositional. Social science can make data science better and data science offers social scientists another set of tools to understand topics of interest to them.

Table 1: Comparing Social Science and Data Science Concepts.

	Social Science	Data Science (DS)
Ontology	Examine relationships between behavioral phenomena and the basis they provide for sharing knowledge across ontologies and methodologies.	DS systems generally operate in numerical rather than symbolic ways, performing statistical inference from large datasets using an iterative optimization procedure that produces effects akin to human learning.
Epistemology	Explain how ideas in the mind could be known to mirror objects/processes outside the mind.	How observable facts can be represented in the memory of a computer for refinement.
Methodology	Objects of interest include humans and human relationships to the world around them. Focus is on theoretically informed fact finding or theory building.	Object of interest is improving the consistency of the algorithm to identify correlation.
Ideology	It varies. Positivist social science seeks to inform theory and tease out and explain social facts using quantitative methods. Humanistic social scientists look for explanation but focus their efforts on understanding different phenomena in different places and times.	Datum and its aggregates arrive from nowhere. They are unbiased factual snippets that mean something probed by increasingly sophisticated algorithms.

The Emergence of Artificial Intelligence

AI was born in the summer of 1956 when leading computer scientists, mathematicians, and linguists came together in a 2-month workshop at Dartmouth College in Hanover, New Hampshire. Their vision was to develop self-learning systems capable of undertaking tasks demanding human-like intelligence. Although the Dartmouth workshop did not yield immediate breakthroughs, it served as a catalyst for networking among key figures in the field. Over the subsequent two decades, these individuals, along with their students and associates at institutions such as MIT, CMU, Stanford, and IBM, would dominate the field (Russell and Norving, 2010). Computers evolved, hardware became smaller, and signals moved more efficiently through processing nodes that increased speed. The computer went from being room-sized to desk, and later, lap-sized. The internet evolved from a secret military computing network to a household appliance. The year 2010 marked a surge in AI development. This growth was fueled by massive increases in computing power, data storage capacity, and improved programming languages and methods. Additionally, a shift in the use of the internet for social media applications and embedded industrial sensors provided access to vast amounts of data, further accelerating progress in the field. Natural language processing (NLP), speech recognition, advanced ML models, computer vision, etc. began to 'build out' what would be generically known as AI.

While powerful, these tools are not without their limitations and potential pitfalls. Bias and Fairness are main critiques about all these approaches (Selbst et al, 2019). Remember, the ideology of data science is that pure discreet facts, when queried en masse, reveal things beyond human capacity. In our view, and that of many others, a primary concern here is the perpetuation and amplification of biases present in the data used for training models. Biases can be encoded in the data and learned by the model, leading to unfair or discriminatory outcomes (see Table 2).

Table 2: Biases in Data Used in Training Models.

Problem	Impact	Citations
Bias and fairness	Perpetuation and amplification of biases present in the data, leading to unfair or discriminatory outcomes.	Selbst et al., 2019
Interpretability and explainability	Difficulty in understanding decisions, hindering trust and acceptance, and challenging error debugging.	Lipton, 2018 Rudin, 2019
Generalization and robustness	Poor performance on unseen data or real-world scenarios, susceptibility to adversarial attacks.	Goodfellow et al., 2014 Szegedy et al., 2013
Data efficiency and sample efficiency	Requirement of large amounts of labeled data and potential for unreliable or biased models.	LeCun et al., 2015

AI and Poverty

Before addressing poverty, let's see how AI found its way to the general field development. The inception of AI's involvement in development can be traced back to the Artificial Intelligence for Development symposium in Stanford, California, in March 2010. As we mentioned before 2010 was the AI boom worldwide. This event marked the beginning of AI's integration into various sectors of development including social, environmental, and healthcare. This symposium generated 749 citations with 30 papers published. The healthcare section had the highest cited papers accompanying with the paper applied data mining technologies to link recorded road characteristics to accident severity in Ethiopia (Beshah and Hill, 2010). Notable contributions in healthcare included utilizing AI for detecting diabetic retinopathy from retinal images Silberman et al (2010).

Following the symposium, there was a notable shift toward exploring the potential of AI in promoting sustainable development, particularly after the introduction of the concept in 2015. Goralski and Tan (2020) examined the question of whether AI would accelerate progress toward achieving the United Nations Sustainable Development Goals (SDGs) or exacerbate economic uncertainty, environmental degradation, and social unrest. They adopted a multidisciplinary approach, drawing insights from both business strategy and public policy to assess the implications of AI on sustainable development, with a specific emphasis on advancing the SDGs.

The discussion in development shifted toward data science, particularly big data analytics and ML models (Hilbert 2016; De-Aarteaga et al. 2018). In his 2016 study, Hilbert examined both the potential for, and obstacles associated with, employing big data analytics in international development. He emphasized the capacity for enhancing decision-making in crucial domains such as healthcare, economic productivity, and security. However, he also addressed pertinent issues such as privacy infringements and resource scarcities prevalent in developing nations. De-Aarteaga et al (2018) examined key themes within existing research, examining how best practices from both ML and development studies could shape successful ML for development (ML4D) initiatives. Furthermore, they proposed that limitations in computation and data, which are often seen as obstacles to ML in the developing world, should be viewed as opportunities for innovative ML research. Researchers, they claimed, should investigate how these challenges can drive the creation of entirely new ML approaches.

More recently, in 2020, the concept of 'AI for Social Good' emerged and advocated for interdisciplinary collaborations between AI researchers and development experts to address social challenges aligned with the SDGs. Tomasev et al (2020) outlined a roadmap for fostering long-term collaboration between AI researchers and domain specialists. By drawing on existing AI for Social Good (AI4SG) projects, it identified crucial areas where future AI applications can contribute meaningfully to social progress.

AI and Poverty, Poverty Identification, and Prediction

In recent years, researchers have used AI to identify and classify poverty. Mhlanga (2020, 2021) explored the influence of AI on diminishing poverty. Findings indicated that AI facilitates poverty reduction by enhancing the gathering of poverty-related data via poverty mapping, transforming agricultural education, and fostering financial inclusion within the finance sector. Moreover, the research uncovered significant AI contributions to education and the financial sector, enabling previously marginalized individuals to engage in the mainstream economy.

Similarly, ML and remote sensing (satellite imagery) are seen as a new way to identify and predict poverty. For Hall et al. (2022) typical measures of poverty (e.g., surveys) are expensive and time-consuming. They note that this latter form of data collection dates back to the 1980s and suffer from, as we noted above, lags between data collection and publication, limited spatial coverage, and other factors. Because of these limiting factors, remote sensing researchers have been developing algorithms to predict poverty based on nighttime luminosity (NTL) (cf, Yeh et al 2020; Blumenstock 2016). With recent developments in solar inflection and convolutional neural networks, using NTL can now assess the relationship between daytime and nighttime. The lighted areas are, for these researchers, indicative of economic activity and therefore able to predict indicators of poverty at the regional level (Jean et al. (2016).

So, correlating nighttime light is an indicator of economic activity? What if economic activity takes place away from one's community or region? Cocoa is a cash crop in the Ivory Coast and Ghana – indeed they produce around 70% of the global supply of cocoa. The cocoa markets are in cities like Abidjan, Cape Coast, or Tema, far from where they were grown. A dark, populated area, then, does not indicate poverty (Carr 2011; Hill 1964). Remote sensing cannot see the weaver or the tailor working by nightlight. It cannot see people spending their money at a chop bar in a rural town in Ghana. And, most importantly, it cannot comprehend economic activity outside the region of focus.

There is value in these threads of research. These tools enable us to identify and visualize concentrations of poverty. Knowing where poverty exists, and its multiple signatures, is critically important in poverty studies. Where these analyses fail is in their capacity for translation into policy and intervention. Our discussion now picks up this discussion around poverty identification and alleviation.

Identifying and Aggregating Poverty Versus Alleviating Poverty

Sen (1976) conceptualized poverty measurement as having two steps: (1) identification and (2) aggregation. For Sen, identification is simply that: who are the poor? Aggregation makes this data legible across society, not between individuals, or among households. Foster et al. (1984) developed 'FGT family of indices' which became the standard for international organizations such as the United Nations, the World Bank, and USAID. The FGT family of indices is a unidimensional poverty index with a focus on income. A key indicator, then, is the 'poverty gap', which identifies and determines poverty based on the distribution of income. Traditional unidimensional measures assess poverty in terms of income distribution as indicated by the poverty line. Proponents of this approach argue that the measurement of poverty is a purely technical process driven by the need for solid and unbiased data collected according to the best practices for household surveys. In general, there is a compromise and trade-off between identification and aggregation: typically, the qualitative view focuses on identification and the quantitative view focuses on aggregation. Typically, governments choose to minimize statistical measurement error at the expense of accepting greater representational error: the perspectives, opinions, and agency of poor people are not part of the equation.

In an effort to overcome this, researchers associated with the Oxford Poverty and Human Development Initiative designed a Multidimensional Poverty Index (MPI). Developed by Alkire and Foster (2011) the MPI moves away from solely examining income distribution. The MPI measures poverty-based deprivation. They ask, and answer, what 'sort of data are appropriate to measure poverty?'

To do this, they propose measuring three dimensions: health, education, and standard of living. They created the index using concepts such as experience of deprivation (e.g., poor health, lack of education, low standards of living, and lack of income), disempowerment, threat of violence, and work activities. The MPI is employed by the UNDP, the World Bank, and other multilateral donors.

In contrast to the unidimensional approaches, the MPI proposes measuring poverty in that it gives representation to poor families and respects their prioritization. In terms of the policy of representation, we should question those narratives that control the definition and measurement of poverty. And how do we evaluate the capacity of a given social order to provide well-being to as many people as possible? The policy of prioritization is a challenge: Which policy priorities are based on agreed definitions? To what extent has the role of poverty in global development been defined (see Table 3)?

Table 3

	Major indexes and tools	**Focus areas**	**Purpose/use**
OECD	Better Life Index	Income, jobs, health, access to education, environmental quality, life satisfaction, safety, etc.	Compare well-being across countries and to better understand what drives well-being in different national contexts.
World Bank	World Development Indicators (WDI)	Education, health, poverty, environment, and economy	Provide a comprehensive set of data about development that allows countries to measure their progress against other nations.
UNDP/ OPHI	Global Multidimensional Poverty Index (MPI)	Health, education, and standard of living	Developed by OPHI and the UNDP, the MPI measures and compares poverty beyond income to multiple deprivations that individuals might face.

Table 3 (continued)

	Major indexes and tools	Focus areas	Purpose/use
USAID	Demographic and Health Surveys (DHS)	Health, population trends, nutrition, and education	To monitor and evaluate the impact of health, nutrition, and population programs.
	Foreign Assistance Dashboard	Economic development, humanitarian assistance, health, democracy, governance, and education	To track and analyze the impact of US foreign aid around the world in various sectors.

A Way Forward: The Poverty Stoplight Approach for Alleviating Poverty

Mainstream economist Jeffrey Sachs has succinctly noted that economic information is terrible for predicting which sectors will grow and innovate. His point is that deductive approaches ignore on-the-ground or localized data because their data sources and indicators seek objective measures and not the measures of the common person. Instead of measuring poverty deductively, like the MPI, the Stoplight measures multidimensional poverty inductively.

The Stoplight has been deployed in hundreds of communities in nearly 60 countries. In Paraguay, where the Stoplight originated with funding from USAID, the government identifies and impoverished person as one who makes \$4/person or \$17/household per day. For the World Bank, the poverty line is ~\$2 per day. Similarly, the United Nations says that people living on less that \$2/day are poor. The Stoplight examines money as a factor in a household's poverty, but also examines another 49 indicators, grouped into six dimensions. The dimensions include health, housing, infrastructure, culture, education, motivation, personal development, and finally, income.

For the Stoplight the unit of measurement is the family/household, which is defined as those who eat and sleep under the same roof. The Stoplight challenges the current paradigm of individualism and fragmentation to account for different norms of 'family' or 'household'. The logic is simple because everyone who lives and sleeps under the same roof shares achievements and hardships. There is no 'poor child' in a home whose mother is 'not poor' or vice versa.

The progenitors of the Stoplight see conventional surveys as 'extractive'. In contrast, the Stoplight methodology employs listening techniques and measurements identified by families. This way the family members can frame their needs and aspirations for the extension agent. Although many responses may be 'subjective', this is a strength. Such responses serve to raise awareness of those who complete their self-survey and to reveal the situation of the poor from the perspective and perception of the poor. For example, governments measure indicators such as financial inclusion or food security. But from the families' perspective, they ask: Do we eat nutritious food at home? Do we have a savings account?

The Stoplight has been implemented in 57 countries and has data on over 500,000 families. In Paraguay, for example, the Stoplight methodology led to the deployment of approximately 6,000 family extension agents across the country at a rate of one for every 150 families. With the extension agent, families reflect on images with simple definitions and rate themselves and mark whether they are green (not poor), yellow (poor), or red (very poor) for each of the 50 indicators; then they prioritize their yellows and reds and develop a plan to overcome their subjective poverty. In other words, this methodology stops using only 'indices' that add statistical information for government decision-makers and starts providing 'dashboards' to women managing a household.

Extension agents support families and, based on their responses, connect them to different public services, or opportunities such as microfinance. Each family has the potential to overcome their poverty class when they are allowed to define and visualize their challenges with accuracy.

Table 4

Families who take the survey are able to:

→ Visualize their own poverty in a simplified way.

→ Become actors, not objects, of development and poverty elimination.

→ Break down the overwhelming concept of poverty into manageable fragments.

→ Develop their own Life Map to overcome poverty.

The Poverty Stoplight does not ignore the structural responsibility of other private and state actors. However, when thinking of community economies, solidarity economies, or diverse economies, the Stoplight highlights the central role of poor people in defining and overcoming their poverty. It is about activating and respecting people's agency and self-efficacy so they can be free to do what they want and achieve the goals they consider important.

Today, computing capacity enables us to examine millions, billions, and trillions of records that would take humans decades to process 'through their senses'.AI is one set of tools that enable researchers to train computers to use statistics and other computational tools without being programmed (Samuel 1957).

The Poverty Stoplight Methodology and Dataset

The Stoplight works because it facilitates a process whereby individuals, families, and communities can better understand the areas where they feel impoverished, not areas that are determined by a think tank, policymakers, or cadre of academics far away. Trained workers collaborate with families and communities to develop their own set of poverty indicators – well beyond the $2.15 per day the World Bank suggests. Then, using an app on a phone, tablet, or computer, families take a visual survey to produce a poverty map that allows them to see the details of their stated needs on a dashboard. Families select images, categorized as red, yellow, or green, that resemble their reality for each poverty indicator (see Figure 2).

Overall, households who participate in taking the Poverty Stoplight survey are able to:

- visualize their 'poverty',
- be agents, not objects, of poverty elimination,
- prioritize needs and set goals, and
- develop their own Life Map.

A family's goal, then, is to upgrade every yellow and red level eventually to green. The data collected from these activities can visualized in a number of ways as demonstrated in Figure 3.

Also in Figure 3, we can see the beginnings of economic opportunity. For example, we see the inability to budget or plan is in the red for almost 40% of people. This is also a basic skill of entrepreneurship. This offers insight into the types of training that can be leveraged to overcome multiple forms of poverty. Related to this we see that 40% of respondents either are unsure or feel they do not have the knowledge or skills to create an income. Part of this capacity comes with understanding opportunity. This brings us to our third example, valuing traditions and culture. Here we can see a concrete business opportunity in the heritage and culture industry. What form this may take depends on the location of the community. It could be making traditional products to sell locally or to tourists farther away. It could be an opportunity for culturally informed eco-tourism, too. These are just examples for one small set of indicators involving 80 respondents. Imagine tens of thousands of respondents, a full set of 50 indicators, and an ML model that runs supervised and/or unsupervised.

Because all 'stoplights' are geo-tagged, the app then generates poverty maps for entire communities, allowing stakeholders to make targeted efforts and to better channel resources in a joint effort to eliminate poverty.

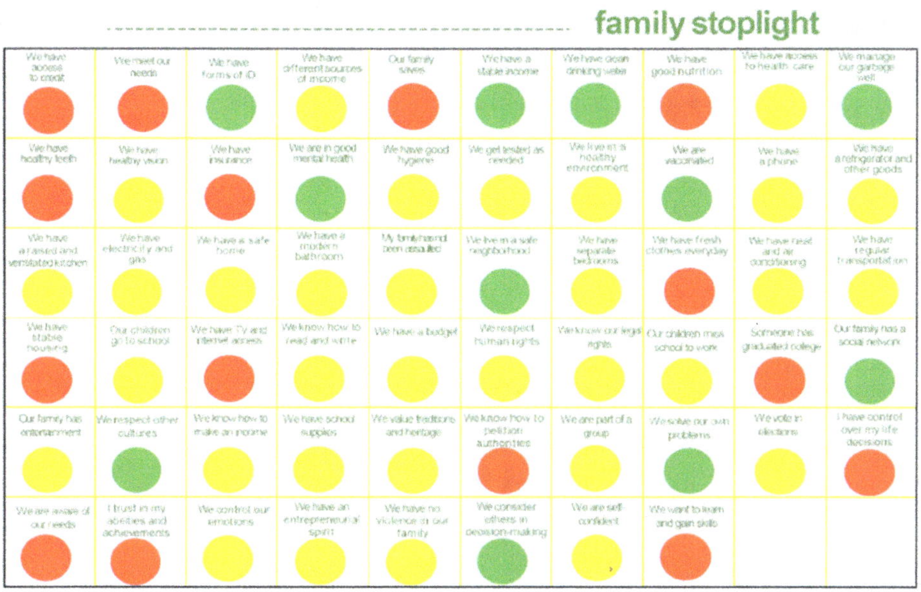

Figure 4: Poverty Stoplight Dashboard[3]

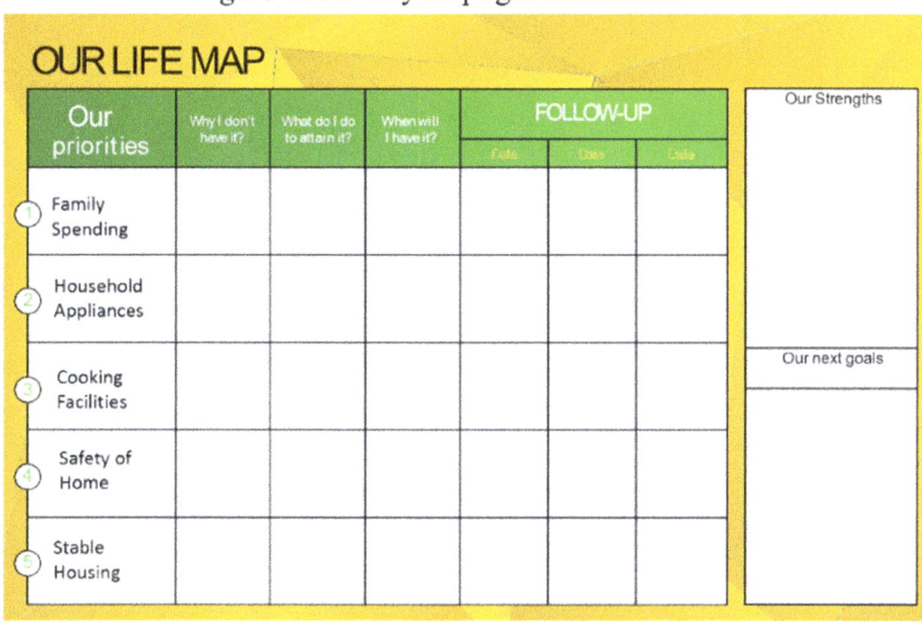

Figure 2

The key takeaway here is that, in contrast to centralized data collecting organizations, the Poverty Stoplight does not assume how people define their own poverty. The basis

0%
10%
20%
30%
40%
50%
60%
70%
80%
90%
Children enrolled and attend school
Access to information
Literacy
Ability to plan and budget
Awareness of human rights
Child labor
College graduation
Communication and social capital
Entertainment and physical activities
Knowledge and skills to generate income
Respect for other cultures
School books and supplies
Value cultural traditions and heritage
Influence in the public sector
Participation in community activities
Conflict resolution
Voting and civic engagement
Autonomy
Knowledge of needs
Self-confidence and self-esteem
Emotional-affective awareness
Entrepreneurship
Family violence
Moral awareness
Willingness / desire to develop skills and knowledge

Figure 3

of the Stoplight is simple. It assists poor families in assessing their own poverty and uses this knowledge as a tool that empowers. In contrast to other approaches, the Stoplight does not lead a participant to believe they are 'poor' because they earn $X/day. Rather, they are offered the opportunity to examine their challenges and the options they possess for changing their lives.

Conclusion

Social scientists use the scientific method to understand social problems. Situated causality can be apprehended by the scientific method. In data science, however, causality is eclipsed by correlation because of the massive number of correlates in the enterprise of big data. For example, Anderson (2008) describes this capacity for correlation extending from the massive amounts, and ever-growing, cloud data that is examined with algorithms that specialize in ML, NLP, prediction, and pattern recognition, to name a few. In short, flawed as it is, when referring to AI the social scientific method gives way to explanation through correlation.

So far, we have examined the 'science' of AI from a social scientific perspective. We have elucidated issues in AI techniques that represent a form of social science that existed two centuries ago. Our intention was to bring a critical eye to the use of AI in poverty identification and assessment.

We have raised concerns about the use of:

- data which may have been captured for other purposes;
- assumptions and perceived norms to determine human behavior; and
- AI in poverty measurement and assessment.

We then refined our focus to poverty assessment as practiced in data science. We discussed its roots and major areas of concern and practice. We examined the critiques of poverty assessment raised by Sen in the 1970s and extrapolated those critiques to modern-day use of AI.

In the course of this discussion, we introduced the Poverty Stoplight, which was developed by coauthor Martín Burt. By way of example, we described how the Poverty Stoplight offered an antidote to the way poverty analysis had traditionally been done as well as through current methods that employ AI. We showed the ways that the Stoplight upheld core principles of social science that were abandoned by data scientists and those who use AI. These inputs have brought us to the current point: that AI is a useful tool for assessing poverty when it maintains the integrity of the practice of interpretivist social science.

Data collected through the Stoplight, and other purposive surveys and questionnaires, could be examined using ML models to reveal a richer picture of families, communities, and their assets rather than their 'impairments'. Indeed, the development and humanitarian communities collect massive amounts of user date annually through program evaluation or Monitoring, Evaluation, and Learning. CommCare, for example, is an open-source application that INGOs and governments use to collect, report, and store data. It can include notes from practitioners, health data of beneficiaries, photos, and even GPS coordinates. DevResults is another package, which is web-based, and serves many of the same functions as CommCare. DELTA is a multisectoral monitoring and evaluation tool with a dashboard that can generate reports. All of these tools are used to collect and store beneficiary data. They rely on simple statistics for their dashboards. These data are often better than the census data or country-wide data collected by aid agencies and governments. They are, however, collected for certain purposes, modes of evaluation, and criteria for program success.

The Stoplight can provide examples of how inductively captured research, examined with ML tools and methods, could support changes that could alleviate poverty. The approach would provide a more granular analysis than currently available through census data. The approach would use practices equated with interpretivist social scientific theories of human behavior and social change. Correlation of proxy data could be replaced with generative data that reveals solutions as well as problems and in the words of the proposed beneficiary.

Berk Calli and Yunus Doğan Telliel

Chapter 8 Environmental Robotics and Its Futures: Extended Temporality of Ethical Engineering

Introduction

The previous chapter examined one tool from the AI toolbox, machine learning. The chapter argued that we need more convergent approaches to understanding poverty, from the perspective of those who are impoverished, to aligning data science tools and social science methodologies, to create better development outcomes. In this chapter, we examine, from a technical perspective, advances in machine learning, sensors, control systems, and other areas that have contributed to robotics. These technologies have made robots more 'intelligent', 'agile', 'robust', and 'resilient'. Consequently, robotics has enabled many new economical and societal development opportunities, and transformed almost every industry, be it manufacturing, health, agriculture, logistics, or defense.

At the same time, robotics introduced various societal challenges, including the effects on existing workforce and future workforce development (Borenstein, 2011), liability complications of autonomous systems (Tai, 2018), racism and other forms of discrimination (Asaro, 2016), and dehumanization of users and their communities (Sharkey and Sharkey, 2012). As such, the need for developing ethical and societal paradigms to guide the robotics development has been acknowledged in the engineering community. Such efforts have been gaining significant visibility in public conversations (IEEE, 2021). Nevertheless, we argue that a specific sub-domain, environmental robotics, requires a special attention and treatment, since its implications, mechanisms, and impact on economic and societal development pose unique challenges and opportunities.

This emerging field of robotics has significant potential to contribute to the current efforts of mitigating climate change's effect on human and nonhuman inhabitants of the planet. The urgency of the climate crises make is crucial to invest in environmental robotics. At the same time, the long-term and challenging-to-measure benefit returns of this technology make it difficult to pursue conventional profit-driven investment strategies. Since local investments can only provide limited outcomes, it is also vital to develop a global approach in the development and utilization of this technology. Considering these factors, this chapter discusses ethical perspectives that can offer a critical understanding of environmental robotics' implications for economic and societal development as well as a set of principles to guide research and development efforts. Approaching 'environmental robotics' as an emerging and promising area of innovation for the public good, the chapter presents a new direc-

https://doi.org/10.1515/9783112218716-010

tion for public interest robotics that aims to center the transformative power of ethical engineering practices.

Since the adoption of ABET's Engineering Criteria 2000 in 1996, most engineers trained in the United States have become familiar with 'ethical engineering' and its primary concerns. Indeed, educators in the US engineering programs have translated Engineering Criteria 2000s emphasis on 'professional and ethical responsibility' into pedagogical practices and curricular offerings that put ethical engineering as a cornerstone of engineering education. The ethical engineering framework is mainly interested in temporary solutions to practical issues that might come up in technology design, delivery, and deployment. What is missing in this framework is an ethics that goes beyond practical moral conflicts and becomes a transformative force that aims to undo the structures of power and inequities in engineering and society as a whole. Unlike the mainstream 'ethical engineering' framework, the alternative would concentrate on the very conditions (structural issues involving classism, racism, sexism, ableism, ageism, etc.) that give rise to moral difficulties and challenges in particular instances. We believe that environmental robotics as an application domain has the potential to develop such a transformative ethics perspective. We are interested in environmental robotics for four main reasons:

First, environmental robotics is well-positioned to respond to issues related to climate change and its impact on communities around the world. As climate change's effects on the planet is in a 'critical zone' (Rockström et al, 2009; Steffen et al., 2015), the adaptation and resiliency of human inhabitants require a radical shift in how engineers and technologists think of sustainability in and through technological systems. As environmental robotics aligns its research and development agenda with the global efforts to mitigate the impact of climate change, it can become both a site and a vehicle for such a shift in 'sustainability thinking' in engineering.

Second, there is a strong affinity between environmental justice and the transformative ethics perspective that we believe environmental robotics can embed. Environmental justice movement is a vital source of advocacy for those who have been disproportionally impacted by pollution and climate change (Sze and London, 2008; Mohai et al., 2009). In order to meaningfully respond to massive disparities in the distribution of the burden of environmental degradation, environmental justice prioritizes advocacy against the uneven distribution of wealth and power. This very sensibility is also central to the cultivation of a transformative ethics in robotics.

Third, environmental robotics' impact is not only environmental, but also societal – more precisely socio-environmental. Environmental robotic technologies can lead to the displacement of workers from their previous jobs, or introduce new work routines and expectations that can be harmful to workers' collective and personal wellbeing. Similar tensions have been discussed by scholars who try to integrate policy-focused 'just transitions' with tech-focused 'design for transition' within the framework of a new 'green deal' (cf. White 2020). In a similar manner, environmental roboticists need to look for ways to navigate across environmental and societal impacts that may diverge from each other. The transformative ethics perspective can help roboticists with this task, so that a positive environmental impact does not have to compete with a worker-friendly impact that can, say, prevent the alienation and marginalization of workers due to the introduction of new technologies.

Lastly, environmental robotics demands a way of thinking that is not closely aligned with the general tendency in tech innovation – that is, to expect a high return for investment in a short amount of time. Environmental robotics might not work well with corporations' goals to serve their own private interest and might need a view of technology as a public good. As such, environmental robotics and its ethical implications need a reconsideration of the mainstream ideas about innovation, investment, and time.

As the present is full of potential futures that robotics research can help actualize and materialize, the past is part of the present as the latter is shaped by social structures, technological infrastructures, cultural viewpoints, and knowledge practices that have been transmitted from past generations. We believe that without a proper acknowledgement of the presence of the past in the present, a future-oriented outlook will not be able to fully capture the (negative or positive) impact of robotic technologies. Scholars in engineering studies have pointed out that contextualization is one of the main methods that help engineers move away from a narrow technical solutionism (cf. Stettler Kleine et al. 2021). We think that the perspective of an extended temporality (the past and future as temporal vectors extended into the present) is not simply a matter of contextualization. Yes, the context is important. Robotics research and development are done by engineers and other technologists who are part of the society, and their knowledge, imagination, and questions do not emerge in a vacuum. Their context matters. But, with the idea of an extended temporality, we are pointing at the possibility of 'a transformative ethics' that requires an understanding of what needs to be transformed – that is the presence of the past in the present – in order to reach a just and equitable future for all.

The chapter is divided into two main sections. The first section introduces the idea of public interest robotics. Public interest technology is a growing movement of researchers, educators, students, and practitioners who want to recenter technology education and innovation on principles of the common good (Abbas, 2021; McGuinness and Schank, 2021; Ward and Bruce 2022). While public interest technology has received scholarly attention in the last years, ours is the first attempt to describe a framework for public interest robotics. The second section moves to a discussion of waste sorting robots as an example of environmental robotics. Our discussion suggests that a future-oriented outlook can benefit from a close attention to local economies and worker communities. In the conclusion, we offer a new perspective that brings the past and the future together as a way of giving life to a transformative ethics in environmental robotics.

Robotics in the Public Interest

As robotics research has matured, there is a growing interest in utilizing robots for international development. From cleaning the oceans to waste recycling and from disaster response to forest management, these technologies can indeed provide a large array of solutions to urgent environmental problems and have a great potential to contribute to the well-being of the society. Yet, environmental applications have not been traditionally a strong focus in the robotics field. Because of the nature of physical spaces environmental robots operate in, such applications require robots to operate in highly cluttered and relatively unstructured environments (such as recycling facilities, scrap cutting yards, or farming fields) that comprise excessively challenging operating conditions. Most robotics researchers believe that, with numerous advances in robotics research and design in recent years, robots are becoming capable of coping with challenges of 'unstructured' settings, enabling new opportunities for the utilization of robots in an expanding range of environmental applications.

There have been an increasing number of research and development efforts in robotics to tackle environmental problems. For instance, regarding atmospheric particle pollution some researchers developed robots monitor air quality (Jiang et al., 2019; Polyakov & Fedyanin, 2020; Rohi et al., 2020); for assessing the biochemical flows, Sadowsky et al. (2006) developed a robotic system to monitor phosphomonoesterase activity in soils; to preserve biodiversity, several researchers designed robots to terminate invasive non-native species (RSE: Robots In Service of the Environment | Lionfish Project, n.d.; Zeldovich, 2019); and regarding freshwater availability, Sepulveda et al. (2020) developed a bio-surveillance robot.

In environmental robotics, there are four main types of solutions: prevention, intervention, mitigation, and adaptation. Prevention solutions focus on stopping or reducing the devastating effects of a potential environmental disaster. Monitoring robots (Sepulveda et al., 2020; Tiwari & Chong, 2019; Vakilian & Massah, 2017) are central examples of this category. Intervention solutions are designed to react to an ongoing disaster whose destructive effects are at or near its peak. In these situations, the disaster is often dynamic and evolving. Forest fire intervention robots (Merino et al., 2012) and oil pipe repairing robots (Shukla & Karki, 2016) are some examples. Mitigation solutions aim to lighten the effects of a disaster after it is under control at large. Robots for reforestation (Zaoui et al., 2015) and for cleaning oil spills (Zahugi et al., 2013) are some examples of this category. Adaptation solutions are for accommodating the changes in the environmental conditions and providing resilience for the communities and nature. It is often difficult to define when the catastrophic effects of an environmental disaster start or when they end, and therefore, it is preferable not to strictly define a developed robotic technology with one of these four categories. Still, these four categories are helpful in providing a bird's eye view of environmental robotics.

Because of the kind of challenges entailed to the deployment of robotic technology in this domain, we believe that environmental robotics needs to entail an ethical framework with which social, economic, and environmental issues are thoroughly analyzed. We also believe that environmental robotics can and should have something more to offer than traditional engineering research. For instance, research focusing on agricultural applications may contribute to the know-how of environmental robotics, but such research would be limited unless it seeks to generate positive environmental impact. An agricultural robotics project focusing on reducing the use of chemical waste or the watering needs would align more with the core ethical values of environmental robotics, while robots that only focus on efficiently picking crops would not. To give another example, a robotics application that has the goal of reducing human footprint on the planet – such as the design of robots to deliver better recycling outcomes, or of an underwater robot to collect ocean waste – would be a good example of environmental robotics while a robotic system that is designed for making an electric car manufacturing process more efficient and profitable would not. This is not to deny that such engineering research and design projects might have positive environmental outcomes (e.g., using fewer resources) and reduce the environmental burden. But, in order to present environmental robotics as a field that integrates a transformative ethics perspective into technological innovation, the articulation of research questions should be directly related to environmental justice and other ethical concerns.

Many industries that have (or are soon expected to have) a strong robotics presence are driven by private incentives in a demand-oriented, profit-driven manner. In the case of environmental robotics, this model is very difficult to establish and sustain since the impact of the technology, or the lack thereof, can mostly be observed in the long term. Environmental robotics' success thus depends on a research, design, and implementation framework that is able to go beyond the mainstream innovation visions that focus on commercialization and private ownership at the expense of civic and public interest. Compared to the investment on industrial robotics and other financially profitable domains, application of robotics in public and civic sectors is significantly limited. How can we then think about a domain of robotics that cannot have short-term consequences or financial returns on the investment? This is an urgent question given that environmental problems are deeply impacting the well-being of individuals and communities. Furthermore, environmental concerns are not evenly distributed across communities. Historically marginalized groups have been more severely impacted by environmental injustices. All these complexities demand a view that is not as limiting and limited as what the private interest wants to see in technology innovation.

We argue that, in order for a domain like environmental robotics to grow, technology innovation needs an intellectual and funding structure transcending private interests of companies. While corporate executives and engineers in corporations might have strong moral values and commitments to the collective good or social justice, the logic of capitalist growth often limit what can be done with technology for the broader social good. Of course, public interest robotics does not have to ignore companies and private entrepreneurial activities. Yet, if robotic technologies are only seen as private goods, it will be very difficult to address environmental injustices. By opening up new spaces of robotics innovation and connecting with new social actors, a public interest robotics framework can help robotics researchers and developers: (a) identify high impact areas, (b) assess the effect of the robotics solutions on different stakeholders, and (c) design robotic solutions that are mindful of unintended consequences.

The public interest robotics framework is especially helpful for advocating research support in environmental robotics. While the progress in engineering sciences has made robotics researchers optimistic about environmental applications, there remains a bigger question about the allocation of resources to support research. Compared to other robotics research domains such as service, medicine, military, or industrial robotics, environmental robotics have been supported by a smaller number of funding opportunities. Furthermore, the relative lack of systematic and coordinated investigation into environmental robotics has been limiting researchers' ability to set up new research agendas that can help bring together disparate research trajectories under a more impactful research framework. This is especially unfortunate because there are already many well-established and well-resourced platforms focusing on environmental problems (from supranational organizations, to governmental agencies, to local municipal initiatives), yet robotics receive little attention in such platforms. Therefore, establishing communication between the public, robotics researchers and policymakers, administering government funded projects, and putting in place suitable tech policies are of key importance to achieve effective solutions through environmental robotics.

Along these lines, with the support of a Public Interest Technology University Network grant, in 2022 we introduced a portal for environmental robotics, www.environmental-robotics.com, which presents past and current research projects in academia, existing companies and start-ups, current and past funding opportunities, and possible future solutions. We plan to leverage this website toward general efforts to establish a community of researchers, educators, and practitioners, and eventually organize interdisciplinary meetings and conference. The main goal is to connect stakeholders across the spectrum of academia, industry, civil society, and the public sector, in such a way that technology development can be integrated into concerns for social justice, environmental sustainability, and economic prosperity.

Like the science and engineering problems discussed in the other chapters of this volume, environmental robotics research inquiries tackle with what are called 'wicked problems'. These issues require an interdisciplinary approach as well as an awareness of broader social implications. Current environmental problems resist one-dimensional solutions. Robotic technologies add another layer of complexity to these situations that already resist straightforward definitions. The deployment of robotic technologies may require new ways of organizing relations of work, expertise, and ownership. Furthermore, the development and design of a robotic technology is only one step in the life of that specific technology. Having resources to train and recruit skilled workers that can operate or maintain or repair a robotic technology is as important as the design and development of such technology. If environmental robotics succeeds in acquiring the recognition and support it needs as a domain of research and development, it will take robots out of the 'controlled' environments of labs, factories, warehouses, and airports to the 'wild'. The contexts within which these robotic technologies operate would be extremely different from each other. Thus, environmental robotics will have to be responsive to the contextual factors of the deployment in order to generate positive impact.

Environmental Robotics for Waste Sorting

In this section, we zoom in on an example of environmental robotics: robotic waste sorting for recycling. Our discussion is not meant to be exhaustive. We want to highlight some contours of environmental robotics with respect to this application. Our goal is to establish a departure point for a new way of thinking about robotics research and its social implications for development. The emerging role of robots in materials recovery facilities (MRFs) provides a representative example for the potential outcomes of robotics in environmental applications. In the United States, the recycling industry represents $117 billion in economic activity and employs more than 530 thousand workers (Institute of Scrap Recycling Industries (ISRI), 2018). Yet, the recycling rates are far from desired levels; under 35% of the waste is actually getting recycled in the United States. MRFs have a critical role in municipal waste recycling. They are the facilities where the collected mixed recycling waste is sorted into separate piles based on the material type, such as paper, plastic, glass, metal, and various subcategories.

Even though a large array of machinery is employed in MRFs, manual work has an integral role in waste sorting. In several stages of the MRF plants, workers are stationed next to conveyor belts, identifying and picking materials from a moving waste pile. This work has many challenges: injuries can occur due to sharp or toxic objects in the waste stream, the working conditions are dirty and uncomfortable, and ergonomic injuries are common in waste sorting workers. Some MRFs report major attrition problems, having challenges to continue their process. The profit margins of MRFs are quite narrow as well. The value of the end-products, the sorted materials, depends on the sorting accuracy, which is often below desired levels, resulting in either small profit for MRFs (e.g., sorted paper is generally worth $10 to $30 per ton), or even some cases they pay to get rid of piles with too much contaminants.

The role of governmental policies and support is also foundational for environmental robotics research and development. The policies would be essential to avoid purely profit-driven solutions, which might be suboptimal in terms of environmental outcomes and social policies supporting workers. For example, a policy was put into place to protect the jobs in MRFs in San Francisco, and to compensate the workers' wages. This policy protects the jobs from the impact of the new and upcoming technology, while creating a case study to analyze the combined human-robot workforce. Similarly, the United States National Science Foundation's Future of Work at the Human-Technology Frontier program has provided funding for investigating new roles for robots and workers that can have an overall positive impact.

This landscape shows that there is much room for improvement in the waste recycling process. Thanks to the recent breakthroughs in the machine learning and robotic manipulation methods, allowing robots to identify and pick waste items in clutter, several robotics companies developed robotic waste sorting solutions (AMP Robotics, n.d.-b; Waste Robotics, n.d.; Tuomas et al., 2014). Interestingly, these solutions came before the research community started looking into this problem, which may also indicate the stagnancy of the robotics research in academy on the environmental applications. Currently, it is too early to comment on the economic and environmental impact of these robots: their performance depends on their operating conditions (mainly the waste stream's level of clutter) and numbers reported by the MRFs about the accuracy of these robots are varying and anecdotal. Therefore, the economic, social, and environmental implications of these technologies are yet to be seen.

In a collaborative project with the participation of Worcester Polytechnic Institute, Yale University, Boston University, and Washington University (NSF Award Search: Award # 1928506 – FW-HTF-RL: Collaborative Research: Shared Autonomy for the Dull, Dirty, and Dangerous: Exploring Division of Labor for Humans and Robots to Transform the Recycling Sorting Industry, n.d.), Berk Calli, one of the authors of this chapter, developed robotics technologies that can maximize environmental benefits and grow the recycling economy. Calli and his team focused on a human-robot collaboration framework aiming to prevent loss or demeaning of work for frontline workers and create an efficient workflow that allows accurate and efficient waste sorting. In this framework, they aimed to combine the strengths of the robots (being fast and resilient) and humans (being intelligent and skillful) to overcome the weaknesses of the overall system. Specifically, they argued that a paradigm that humans and robots can teach each other and cover for each other's mistakes would result in high-quality MRF output.

We call this a derivative robotics solution, as their project starts with an existing, human-worker-based solution and aim to make it more efficient, safer, and profitable while not changing the overall workflow of the process significantly. Similar to the recycling robot discussed in AMP Robotics (n.d.-a), this project aimed to develop a technology that redefines the human's role in a materials recovery facility, while keeping the whole waste sorting process the same in essence. In addition, to promote research in the waste recycling domain Calli and his colleagues published a waste detection and segmentation dataset (Bashkirova et al., 2022), and organized a waste recognition competition (VisDA2022: Visual Domain Adaptation Challenge, n.d.). The dataset is free, public, and the first of its kind, being collected in real MRFs and therefore presents realistic challenges of this domain.

Considering that there are various potential robotics technologies that can be developed for this application, and that each type of technology would have different environmental, occupational, and economical effects, this project adopted a scenario analysis framework (Alcamo (ed.), 2008) to analyze potential outcomes and identify the particular way in which this technology should be developed and implemented. We believe that such transdisciplinary collaboration is essential to develop positive impact solutions for the recycling domain and other environmental robotics applications.

While these technologies can have a significant role in solving the recycling problem, it is important to note that this domain is very complex. As such, the recycling policies and practices differ by country, state, and city. Waste characteristics and challenges of MRFs largely vary based on the location. For instance, think of two different cities in the Northeast United States: New York City, New York and Rutland, and Vermont. Since the consumption and recycling practices of the two cities vary drastically, MRFs in these cities handle different recyclables, contaminant types, and clutter. The waste amounts that need to be processed are also dramatically different, as well as are the size and resources of the MRFs and their worker availability. Even weather conditions play a role: in rainy days, for instance, processing certain recyclables such as paper and cardboard is more challenging. There are also other challenges including a highly fluctuating recycling market, and monthly and even daily waste characteristic changes due to seasons, national holidays, public events, and other factors shaping consumption patterns.

The diverse challenges of MRFs call for diversity in the technology front. Technologies that promote human-robot collaboration can also play a role, in which the robots can be trained by workers and their role can be adapted based on the requirements of the day. Such a technology has a potential to achieve the agility that is required from these systems due to the constantly changing waste characteristics. This scheme also implies new roles for human workers, and needs to be carefully designed to not demean but enhance the meaning of recycling work, which requires a thorough understanding and analysis of workers' perspective (Wrzesniewski, 2003).

Furthermore, an MRF with robotic integration can help enhance the safety of the sorting work. The manual stages of the sorting process are made up of two main stages: a pre-sorting (first) stage and a (secondary) sorting stage (also called quality control). On the one hand, the first stage is more dangerous for workers, but much harder to automate via robotics solutions. On the other hand, the secondary sorting stage is relatively safer, but easier to automate via robots. The solution in the market focuses on the safer secondary sorting stage, and there is practically no derivative or transformative solution proposed for the pre-sorting stage. More public incentives are needed to develop robotic solutions for the pre-sorting phase that can help establish safer working conditions.

To achieve a positive environmental and societal impact in this domain, the landscape of the problem should be understood together with the needs of the various stakeholders. Introducing a new technology to this domain should aim to improve recycling efficiency, and at the same time should take into account its impact on the recycling economy and workers and their communities. If new roles are envisioned for workers within novel human-robot collaboration schemes, the meaning of work should be preserved or – if possible – enhanced. New robotic solutions should aim to improve the safety of manual work in these workplaces. In addition, strategies to avoid or minimize job displacement should be constituting a central concern in the deployment process.

Epilogue

In this chapter, we argued the significant potential of environmental robotics to foster economic and societal development, while discussing challenges to realize such solutions as well as the potential inhibiting effects of this technology domain. Some of these challenges and inhibiting factors are unique to environmental robotics. In many other domains of robotics, for instance, there are similar concerns about machine takeover of various manual jobs and displacement of workers whose voices are not represented in decision-making mechanisms. However, we argued that certain aspects distinguish environmental robotics, namely the urgent need of such technology, its longer-term benefit returns, and the need of applying it to diverse geographical locations. As such, we advocated for a new 'ethical engineering' perspective that is centered on the public interest. Along similar veins, we would like to conclude the chapter by framing environmental robotics from the perspective of the recent studies regarding the United Nations Sustainable Development Goals (SDGs) (Guenat et al., 2022; Mai et al. 2022).

Haidegger et al. (2023) conducted a consensus-based expert elicitation process to identify the positive and negative implications of the robotics technology for achieving the SDGs. The experts stated that robotics would have positive impact for 53% of the environment-related SDGs, while inhibiting the targets for 19% of the overall SDGs. Specifically, robots can be very effective in meeting the targets in SDG 13 (climate action) by monitoring the effects of climate change, and implementing mitigation, adaptation, and resilience strategies. For SDG 15 (life on land), robotics could play an essential role for the detecting unauthorized hunting, mining, and logging, allowing the enforcement of environment protection laws. Similarly, regarding SDG 14 (life below water), ocean robots can help enforce laws regarding the underwater habitat, preventing unsustainable fishing and industrial contamination. Considering SDG

6 (monitor water quality and contamination), robots can help inspect and maintain clean water infrastructure, and reduce contamination by collecting litter. The same expert study also discusses the negative implications and inhibiting factors for a large array of SDGs, e.g., displacing workers (SDG 8 – decent work and economic growth), creating inequality in education (SDG 4 – quality education), decreasing the job opportunities for disadvantaged communities (SDG 10 – reduced inequalities), negative environmental footprint of robotics (SDG 12 – responsible consumption and production), and the potential of enhancing the gender inequality (SDG 5 – gender equality).

So, what is the future of public interest robotics? What does it mean to think about environmental justice with respect to the 'future' of robots? Who does belong to the future that robots enable for us? The 'future' that robotics represents is not a neutral space. For instance, Neda Atanasoski and Kalinda Vora (2019) show that robot-inspired futurism of popular culture and corporate tech development has often been translated into a posthuman fantasy of a life that is free of the human. Yet, this posthuman fantasy has rarely meant the total transformation of the humanity. It has become a coded way of claiming either that some humans (white, male, able-bodied folks) are better equipped to transcend the human or that 'others' will be incorporated in technological systems for the rest to enjoy the benefits of a posthuman world. Indeed, what inequalities and hierarchies will the future of robots create as it 'disrupts' our present-day practices? This question is especially important for us. The interest in and demand for environmental robotics are growing, and we see this as an exciting development. But, in order to prevent environmental robotics from becoming a negative impact on human workers and their communities, we need to align environmental robotics to ongoing efforts to build public interest technology ecosystems.

Technological optimism informs most current conversations on the future of robots. It often wants us to assume that the future happens when the past is over. The allure of the future is so strong that the past seems something one can easily dispose of. It is true that the past can be a burden or an obstacle to innovation and experimentation. But, robots – as socio-technical systems designed, implemented, and maintained by humans – reproduce biases, prejudices, and values that are transmitted from the past. Robots can acquire 'race' or 'gender' because of the way in which they are designed (physical characteristics such as the appearance or voice) or are received by the users (Sparrow 2020; also see a discussion of robotic exoskeleton designed primarily for male body in Rayssiguie and Erden, 2022). Robots can also contribute to already-existing social hierarchies, for instance, by decreasing workers' control on their workspaces and workflows.

Whatever 'future' is premised in the future of robots will be neither a neutral space (where benefits of robotic technologies are equitably distributed), nor totally disconnected from the past. To acknowledge such entanglements is not to deny that robotics is one of the most future-oriented fields of research and design of our times, but to suggest that robotics is like a Janus-faced god: one face looking at the future while the other at the past. Not one or the other, but both at the same time. We want to end this chapter by highlighting this dual orientation. This is, in our view, the biggest ethical challenge facing robotics researchers. Yet, it is also the very condition of building a transformative ethics of robotics research.

The urgency of environmental concerns further complicates the methodologies of developing environmental robotics. Developing an ethical commitment in robotics research, we argue, is a matter of navigating between different temporal horizons. In a society with entrenched social and economic inequalities, being able to create a robotic system whose goal is to contribute to collective and personal well-being requires an awareness of how already-existing hierarchies and power structures are the conditions that shape the problem of wellbeing itself. This is where the past of social structures and ideologies and the future of techno-imagination intersect. And, that is why the past is not what comes before the present, but is instead what lies between the present-day engineering researcher and robots of the future.

The push for development of better (faster, more intelligent, more accurate, more precise, cheaper, more resilient, and more energy-efficient) robots is what fuels the development in robotics research. One of the things robotics researchers are expected to do is to imagine the future of robotic technologies with respect to newer and better computer vision capabilities (to derive meaningful information from visual inputs such as images or videos) or highly integrated sensor-actuator-controller units (to allow for more resilient and faster robots). In what ways can and should robotics researchers leverage such technological innovations – that is, what makes robotics a rapidly growing and increasingly fascinating field – for something other than more efficient and effective robotics systems? We think that environmental robotics has the potential to harness the power of technological innovation to contribute to collective and personal well-being. This is the key to a transformative ethics perspective: redirecting robotics research and development toward building a just world.

Part IV

None of the concepts and practices identified in Parts I, II, and III will be possible without decentering the education experience. We need to 'bake' into education a decentered approach when deploying scientific and technological interventions in low- and middle-income communities.

In Chapter 9, Elizabeth Long Lingo asks if we can reimagine the role of the engineer from singular expert to a leader facilitating a co-creation process among many. Lingo thinks that this is possible with an integrative engineering leadership model. Three main characteristics of this model include an expansive view of networks in the engineering context, emphatic inquiry and co-design with stakeholders, and ability to manage ambiguity. Chapter 9 connects creative leadership concerns to the need for a development engineering that can navigate dynamic environmental, political, and socioeconomic constraints and opportunities.

In Chapter 10, Juan Lucena also questions traditional engineering frameworks. Lucena's focus is on engineering work with NGOs. He suggests that working with NGOs will enable engineers to have more sustained engagements with vulnerable communities in developmental contexts. This is especially important because many engineering students now have opportunities to do community-based work as part of their education, yet this type of engagement is typically temporary. Lucena adds that engineers' NGO careers should have the goal of empowering communities rather than creating vulnerable communities' dependence on external experts.

In Chapter 11, Mimi Sheller, with a focus on her work in Haiti, argues that in post-disaster situations engineers need to be able to adapt to complex social interactions and shifting stakeholders. She suggests that for engineers to be prepared for such situations the pedagogical emphasis in engineering education should be put on interdisciplinary project-based learning that involves participatory research and co-design processes.

https://doi.org/10.1515/9783112218716-011

Elizabeth Long Lingo

Chapter 9 Integrative Engineering Leadership for Sustainable Development

> The heat source is a pair of headlights. A car door alarm signals emergencies. An auto air filter and fan provide climate control. But this contraption has nothing to do with transportation. It is a sturdy, low-cost incubator, designed to keep vulnerable newborns warm during the first fragile days of life. Unlike the notoriously high-maintenance incubators found in neonatal intensive care units in the United States, it is easily repaired, because all of its operational parts come from cars. And while incubators can cost $40,000 or more, this one can be built for less than $1,000. ~Madeline Drexler, "Looking Under the Hood and Seeing an Incubator," The New York Times, December 15, 2008

Introduction

It was a compelling news story – in contrast to the thousands of foreign-made, high-tech, costly incubators that were rendered useless by electricity surges, broken knobs, and lack of training manuals – engineers had designed an incubator for local usability and sustainability. Beyond its news value, though, the resulting headline also hints at a radically different model of leadership enacted by the project's leaders – integrative engineering leadership – one that embraces co-creation with an array of local partners and technical experts. Where previous chapters have focused on broad thinking, techniques, methods, and the like, this chapter focuses on developing creative leadership for co-creating solutions to development problems.

Think about how often you have heard the complaint from colleagues or your students: "I just don't understand it. My idea was amazing! But all I got was resistance when I tried to implement it. Even among people who I thought would be on board – the people who I thought would benefit from it!!" Their complaint arises from the tacit model of leadership that they bring to their work. For example, when many engineers and technologists think of leadership, they invoke the imagination of the sole charismatic visionary who brings forth their technical genius to transform the lives of others, or the brilliant entrepreneur who through sheer determination and grit bring their engineering innovations to market. Many engineering and technology innovation programs implicitly or explicitly reinforce this conception of leadership. However, for those seeking to lead engineering ventures for sustainable development or those who teach those future leaders, such a conception is not only misguided, but can result in abandoned projects, undermined relationships, and unintended negative impacts on local communities.

https://doi.org/10.1515/9783112218716-012

In this chapter, I bring together the latest research on creative leadership and co-creating across networks to present a model of integrative engineering leadership for sustainable development. More specifically, I illuminate three core features of integrative engineering leadership that are often missed in traditional approaches: (1) adopting an expansive network view – situating one's work within a broad network of stakeholders who often have competing ideas, perspectives and interests, yet whose ongoing commitment and passion is needed to ensure implementation and ongoing local sustainability; (2) embracing empathic inquiry and co-creation – rather than technical expertise – as key drivers of leadership success; and (3) recognizing and deftly managing multiple forms of ambiguity throughout the creative process. After discussing each of these facets in detail with illustrative examples, I highlight challenges and opportunities for implementing this model in the classroom.

Models of Creative Leadership

What do environmental engineers developing global climate-change solutions, biomedical engineers seeking to create life-saving devices for low-resource environments, and product engineers seeking to develop low-cost, temporary shelters for climate and political refugees have in common? They are developing solutions that require curiosity, empathy, and facility in co-creating with others. Research on creative leadership illuminates the unique challenges of eliciting and guiding others' creativity, contributions, and passion amidst the ambiguity and roadblocks that are an inevitable part of developing novel solutions and innovative outcomes, and the best practices for doing so.

In their foundational synthesis of the emerging creative leadership field, Mainemelis, Kark, and Epitropaki (2015) identified three creative leadership contexts – facilitative (enhancing and facilitating others' creativity), directive (directing others' actions to advance the leaders' creative vision), and integrative, where leaders both elicit and integrate others' creative contributions, including their own. Engineering leadership for sustainable development, which requires eliciting and integrating diverse creative contributions from both technical and local experts to create long-term viable solutions, is an excellent example of a context requiring integrative creative leadership. The authors highlight the importance of fit – understanding which type of creative leadership context you are in and making sure your leadership approach aligns with that context.

What are the implications of not embracing the right model when engineering for sustainable development? While some creative leadership efforts do require a directive approach (e.g., when implementing a creative solution in an emergency situation), engineering leaders engaged in directive creative leadership are often blindsided by either overt or passive resistance to their efforts. With their primary focus on having others implement their solution or vision, directive creative leaders often fail to recognize the expertise of others and engage in the careful listening and trust-building needed to develop implementable and sustainable ideas with local actors. They assume actors' commitment to the project – especially local experts and gatekeepers – and are often blindsided by those actors' reluctance to implement the leaders' 'good' ideas. Internal team members, sensing the marginalization of their expertise and contributions, are also more likely to disengage – especially as the project meets inevitable challenges and roadblocks.

On the other hand, engineering leaders engaged in solely facilitating others' creativity may excel at eliciting diverse creative contributions and engendering deep trust and enthusiasm among the array of actors involved. However, they often find it hard to align and make tradeoffs across others' competing ideas and interests towards a viable, collective outcome. Concerned about discouraging others or over-exerting their own perspective and vision, facilitative leaders often avoid the conflict and prioritization needed to move ideas forward. This can result in lost momentum and disengagement as actors grow frustrated with lack of progress and perceived fairness regarding how difficult decisions are made.

Key Features of Integrative Creative Engineering Leadership

Integrative creative leadership contexts differ in important ways from creative leadership focused on directing others to advance one's own creative ideas or facilitating the creativity of others. Three core features undergird integrative creative leadership: (1) adopting an expansive network view – situating one's work within a broad network of stakeholders who often have competing ideas, perspectives and interests, yet whose ongoing commitment and passion is needed to ensure implementation and ongoing local sustainability; (2) elevating empathic inquiry and co-creation – rather than technical expertise – as key drivers of leadership success; and (3) recognizing and deftly managing multiple forms of ambiguity arising throughout the creative process. I discuss each of these core dimensions below.

Thinking Expansively about the Project Network

Integrative creative leadership involves eliciting and guiding diverse creative ideas and contributions across a network of disciplinary and local experts towards a novel solution or innovation. Leaders must lead both their immediate project team members, who are immediately involved with the creative process, and those in the broader project network, who may be less involved in the everyday creative process but are nonetheless critical to overall project success (Lundin et al., 2015). Leaders must think expansively not only about the technical experts who are immediately involved in product design and development, but also those whose have invaluable local expertise and knowledge regarding how the product might be used in practice or who may become blockers if not adequately included in the process. Leaders must take a long view of the engineering process to include not only those who can inform product design and development but also those who have a stake or interest in whether and how the project will be implemented and sustained over time (Lingo 2020). Perhaps surprisingly, this more expansive network should include such unexpected actors as suppliers, manufacturers, purchasing agents, repair technicians, delivery truck drivers, electrical system repair personnel – the broad constellation of actors who will ultimately produce and sustain the idea after the team has moved on. For example, as Timothy Prestero, founder and CEO of Design that Matters, discovered, not taking an expansive view of the project network can lead to abandoned projects. He describes how he partnered with local doctors and nurses to develop an award-winning infant incubator that was recognized for its usability in the local low-resource context. However, despite the years dedicated to thoughtful collaboration with local partners, the incubator was ultimately not utilized in their hospitals. Hospital purchasing agents and local medical device manufacturers – key stakeholders who were not considered during the design process – did not have incentive to buy it. As a result, the brilliant technology was never adopted.

In addition to taking an expansive view of the network of stakeholders, integrative creative leaders need to map out the larger context – organizational, cultural, political, legal, historical and technological – in which they seek to advance their innovation (Owens, 2011). Paying attention to context involves assessing how solutions might affect the social fabric and power dynamics among those in the community. Have others tried to solve the problem before? Why or why not, and what was the result? Who might try to subvert the ongoing maintenance of the engineered solution to protect the status quo? In low-resource contexts, the instability of local technological infrastructure may precipitate building redundancies into your solution or developing supportive infrastructure to complement existing technology. For example, in his efforts to help victims who had lost limbs to warfare in the Nuba Mountains of Sudan, Not Impossible's Mick Ebeling not only developed prosthetics that enabled individuals to eat and work independently, but he also set up what is likely the world's first 3D-printing prosthetic lab and training facility and developed the ecosystem to provide materials and energy for the facility over the long term.

Taken together, integrative engineering leadership requires taking a long view of the creative process and the array of actors involved. Questions that help guide integrative engineering leaders in analyzing their context include:

1. What is the history around this problem? What has been tried before? What has worked and did not work, and why?

2. What are the potential organizational, cultural, political, legal, historical, and technological constraints that the project might face?
3. Who is necessary to help the idea come to life (e.g., technical and local experts, users, and resource gatekeepers)? Who might provide unexpected sources of expertise and knowledge?
4. Who comprises the larger network of actors in the project network: government, community members, competitors, accrediting agencies, suppliers, funders, manufacturers, etc.?
5. Who has the expertise and local knowledge to help implement and sustain the innovation over time?

Elevating Empathic Inquiry and Co-creation Rather than Technical Expertise

In addition to adopting a network perspective, leading the design of sustainable solutions in low resource environments requires deep knowledge of local resources, conditions, customs, language, and norms on how work gets done. While a basic understanding of customs and language can be garnered through background research and cross-cultural deep dives, engineering leaders need to engage more deeply to move beyond the espoused ways of doing things to truly understand how things work in the specific local context. For engineering leaders, co-creating with local partners is essential – as these local experts can articulate, illuminate and translate hidden assumptions, constraints and opportunities – for the immediate engineering team and vis-à-vis other actors within the network.

Toward this end, integrative creative leaders embrace the practice of co-creating with local partners throughout the creative process – from initial needs finding to problem definition, ideation, prototyping, testing, further refinement, implementation and ongoing maintenance. Local co-creators provide deep insights into design constraints, previously hidden resources, possible points of resistance and opportunities to re-imagine the impossible. Co-creation is inherently relational and grounded in empathic inquiry, curiosity and humility. It requires valuing both technical expertise and deep insight into social dynamics and local customs.

Integrative creative leadership involves more than generating ideas – it involves an ongoing interplay between collectively generating creative ideas and synthesizing those ideas into an implementable, cohesive whole (Harvey, 2014). Integration demands subtle negotiation, and leaders must strategically finesse how they elicit, share, edit, and winnow ideas while also maintaining the creative energy and commitment necessary to co-create and implement novel outcomes. To do so, an essential aspect of co-creation is to engage the broader array of stakeholders in empathic interviewing to listen and learn. The goal is to engage potential collaborators, funders, and a range of users in a way that enables them to share their needs, interests, knowledge about local opportunities and constraints, and potential points of resistance.

Empathic inquiry is especially important during early stages of the process – to help guide constraint articulation and the possibility space for ideation. However, taking time to engage in empathic interviewing early on is often difficult for many engineers, who have been trained to jump quickly to task advancement and solutioning on a given specific problem. In contrast, it may take time to secure interviews and build sufficient trust with partners to elicit previously unarticulated assumptions and points of resistance. The process of drawing insights from interview data can be messy, demand inductive rather than deductive analysis, and require intentional suspension of assumptions to allow surprising and unanticipated insights to emerge. Engineers may need training in how to engage in empathic interviews – eliciting not only superficial responses and needs but also the subtle reasons 'why' behind other's needs and points of resistance. Nevertheless, engaging in empathic interviewing with key stakeholders to identify unarticulated needs and understand the problem space is critical – saving time and energy over the long arc of the creative process.

Empathic interviewing and co-creating within the larger project network involve more than simply 'pitching' the problem or solution and then listening to how others respond. Interviews also offer the opportunity to engage partners in actively co-creating possible solutions, to build upon each other's ideas in the moment, and use these ideas to further tease out possibilities and possible red flags and constraints. Including local partners in the ongoing co-creation process can also help ease implementation as partners feel ownership of the co-created solution. Local unmet needs and potential red flags can be shared throughout the emergent design process, enabling leaders to negotiate tradeoffs and utilize points of resistance as pragmatic design constraints. Consequently, rather than having to 'pitch' solutions and overcome resistance later on, integrative creative leaders have deftly addressed potential issues already – negotiating integrative solutions that meet the array of needs, perspectives, and interests of various actors.

In addition to co-creating with the broader network, integrative creative leaders also need to pay attention to co-ideation and synthesis within their immediate project team. Co-creating can be problematic if leaders do not pay careful attention to the status dynamics among their team members. Status differences may be subtle – a result of differences in disciplinary backgrounds, roles and titles, or cultural assumptions. To address this issue, integrative leaders must pay close attention to managing status vis-à-vis the array of actors involved in the project. For example, integrative creative leaders carefully craft how they introduce actors to each other – explicitly establishing each actor's expertise and potential contributions to the project as they bring previously unconnected actors together. Further, when giving feedback, integrative leaders do so in a way that recognizes others' expertise and maintains the status of team member's vis-à-vis each other (Lingo and O'Mahony, 2010).

When working across cultural boundaries, leaders also need to attend to cross-cultural differences in behaviors and assumptions related to fostering co-creation. For example, partners may have very different preferences and norms around key aspects of co-creation, including risk-taking and fear-of-failure, providing critical feedback to others, generating possibilities on the fly, and taking turns sharing out ideas (Leung and Wang, 2015). Leaders must explicitly make sure that everyone's voices are elicited, especially when cultural norms may shape when and to what extent team members share their ideas vis-à-vis each other. For example, team members who tend to talk more are more likely to mistakenly equate others' 'airtime' with their knowledge of, or commitment to the project, when in fact it may simply be a result of cultural differences. As another example, dysfunctional group information processing dynamics, such as the common information effect (Gigone and Hastie, 1993), can lead groups to recognize and prioritize commonly held information more than uniquely held information. In turn, individuals who share commonly held knowledge are seen as more competent, compared to those sharing unique information. To mitigate this unproductive dynamic, integrative leaders must actively elicit and highlight the value of diverse and uniquely held information to foster true co-creation.

Fostering co-creation involves not only eliciting ideas, but also helping others to broker and build on each other's ideas to generate new insights and possibilities (Hargadon and Sutton, 1997). Groups often fare poorly compared to individuals in brainstorming tasks due to perceived status differences, the tendency towards group think, and a lack of psychological safety that inhibits individuals' willingness to share 'wild' ideas or works in progress that are invaluable for producing novel solutions (Edmondson, 1999). To counteract these dynamics, leaders must also establish ground rules for respectful multidisciplinary interaction and create an environment that encourages risk-taking, mistake sharing, and building on bold ideas (Sutton and Hargadon, 1996). However, these ground rules may be difficult to achieve in groups with significant differences in status, expertise, and language. In these cases, leaders need to take a stronger hand in guiding co-creation processes. One promising approach that I have used in my classroom is to allow co-creators to brainstorm individually and then ask each party to share out one-by-one so that all voices are heard. Only after everyone has had their ideas teased out (and ideally captured on a white board or common visual space for all to see), do leaders then invite everyone to begin to build upon and combine each other's ideas towards a creative whole.

Elevating empathic inquiry and co-creation – rather than technical expertise – as drivers of innovation are the key to integrative engineering leaders' success. Questions that help guide integrative engineering leaders in this work include:

1. What do the array of stakeholders care about? What are their unmet needs and interests, and why?
2. What might their points of resistance be? Why? What might foster their willingness to adopt and maintain this innovation?
3. Who might be unexpected allies? Unexpected blockers? Why?

4. What potential status differences exist across the array of co-creators? How can everyone be established as an equal expert?
5. How can group processes be managed to make sure everyone's voice is heard, to foster respectful interaction, and to encourage people to build on each other's ideas?

Recognizing and Deftly Managing Multiple Forms of Ambiguity

When many engineers think of 'ambiguity', they often think of open-ended engineering problems where the solution remains open to discovery. However, this is only one facet of the ambiguity faced by integrative engineering leaders. To effectively recognize and deftly manage ambiguity in the creative projects, leaders need a more expansive and subtle understanding of how ambiguity might feature in their work.

Unlike uncertainty, ambiguity is not resolved by more information (Weick, 1995). As the organizational behavior expert and social psychologist Karl Weick writes, "The problem in ambiguity is not that the real world is imperfectly understood and that more information will remedy that. The problem is that information may not resolve misunderstandings" (1995, p. 92). In fact, what makes ambiguity so salient to creative work is that it captures how actors may have very differing perceptions of the same thing and that all of these interpretations can be subjectively plausible and correct. The key question regarding ambiguity then, is not to eliminate it, but how to navigate these differences in productive ways. While ambiguity can be used strategically to allow for buy-in and idea generation, as detailed below, it can also complicate co-creation efforts (Feldman, 1991; Eisenberg, 1984; Weick, 1995).

More specifically, as summarized in Figure 1, there are three primary types of ambiguity that leaders will encounter during the co-creation process: (1) differing perceptions of what constitutes success or quality; (2) whose expertise is most salient; and (3) how the work should proceed (Lingo and O'Mahony, 2010). When not handled deftly, these forms of ambiguity can undermine the quality of solutions generated and leaders' ability to garner and maintain the support needed to advance their projects.

Ambiguity becomes particularly problematic as leaders guide the process of editing and integrating ideas – while maintaining the commitment of everyone involved. It is during these times when disagreements over what constitutes quality or success comes to the fore. When leaders have to say 'no' to others' contributions, or actors have to respond to others' critical (and often not constructive) feedback, deftly managing these differences becomes most salient. To do so effectively, leaders need experience in building trust among uncommon partners, finding common ground, developing shared understanding across competing perspectives, identifying tradeoffs, and

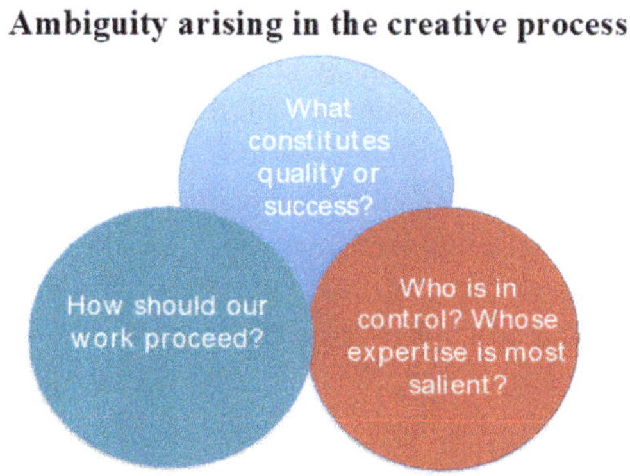

Figure 1: Ambiguity arising in the creative process.

negotiating the differing interests, goals, and values of all contributors to the projects – in a way that maintains the commitment required for implementation.

The challenge of navigating competing perspectives over who should control specific decisions or whose expertise should prevail only complicates these dynamics. However, this may be particularly daunting as integrative creative leaders often lack formal authority over the array of actors involved. Leaders must pay close attention to fostering the passion and commitment of others over time. In addition, stakeholders' roles are often fluid and emergent as the project evolves, with some actors taking a lead initially and then taking a more supportive role as the idea progresses. Thus, integrative leaders play a central role in helping experts carve out these roles as the venture evolves (Lingo and O'Mahony, 2010). Further, in complex global projects, not all contributors are involved in all aspects of the creative process. Leaders thus must also navigate the fluid and evolving involvement of parties over time, effectively bringing newly engaged actors up to speed on the unfolding process and finding ways to elicit actors' input, even when the emergent idea is more developed and refined.

The third form of ambiguity, related to how the work should proceed, further complicates this effort. The most successful integrative creative leaders are savvy at keeping key external stakeholders in the loop – checking in regularly to make sure they are apprised of progress, if not the fine-grained details of the creative process itself (Lingo and O'Mahony, 2010). While ideally, the path forward toward implementation is linear and without problems, there always seem to be inevitable unknown issues. Further, various stakeholders may want to see immediate proof of viability and may perceive time spent on empathic interviewing and discovery as nonproductive. To manage these competing perceptions, integrative creative leaders need to develop shared understanding and commitment to the user-centered design process among all stakeholders – including sharing why such an approach is more efficient and effective in the long run.

Further, leaders can use process to their advantage in managing ambiguity. For example, to avoid costly mistakes, leaders can design and implement small pilot projects and engage in rapid cycles of feedback and improvement – both of which also help to bring reluctant actors on board. Leaders can embrace the power of the 'pivot' – responding to feedback from pilots or changes in the innovation landscape, and communicating the reasons 'why' to all stakeholders involved. Scholars who study entrepreneurs have found that the capacity to pivot not only influences one's ability to attract needed resources, but also and perhaps most critically, impacts individuals' commitment to their venture over the long term (Grimes, 2018). Thus, while leading a nonlinear process is often not typical for engineers, developing the capacity to do so is a key differentiating feature of integrative creative leaders.

Taken together, integrative engineering leaders are keenly aware of how the three types of ambiguity may arise throughout the creative process and proactively work to limit its potentially undermining effects. Questions that help integrative engineering leaders deftly manage ambiguity include:

1. What different assumptions do actors have regarding what constitutes quality and success? What is the 'why' behind their thinking? When and how might these differences become problematic?
2. What different assumptions do actors have regarding who should have control over certain decisions or whose expertise should prevail? What is the 'why' behind their thinking? When and how might these differences become problematic?
3. What different assumptions do actors have regarding how the work should proceed? What is the 'why' behind their thinking? When and how might these differences become problematic?
4. Has our team paused to reflect and pivot as needed?
5. What small-scale pilots could be used to gain insight and to help overcome resistance?

Taken together, integrative engineering leaders are the nexus of co-creation within a network of technical expertise, needed resources, and local knowledge. For many engineering students, this is a radically different conception of leadership. In the next section, I draw from experience in developing hybrid design thinking and engineering leadership curricula to explore how we might provide immersive opportunities to gain greater fluency in this different form of leadership.

Making It Happen: Piloting a Project-Based Learning Experience to Build Integrative Engineering Leaders

At Worcester Polytechnic Institute (WPI), we piloted a new curricular approach to helping engineering students radically experience integrative creative leadership – the transdisciplinary major qualifying project (MQP).

The pilot was initiated by Kris Billiar, Professor and Chair of the Biomedical Engineering (BME) department, who had observed two fundamental challenges in the existing BME curriculum. First, BME students often lacked experience in engaging local users in the needs-finding and problem identification process – problems were typically 'given' at the start of their projects. Second, even when BME students did engage local users, they often developed solutions and brought in the larger network perspective at the end (e.g., to add on a strategic marketing plan or financial analysis). Instead, Professor Billiar sought a project-based learning experience that would truly offer an immersion in co-creation with local users and an expanded network of stakeholders (e.g., government agencies, suppliers, marketers, manufacturers, distribution, purchasing agents, and employers.). In addition, the pilot was co-created with Rob Krueger, Professor and Chair of the Social Science and Policy Studies department, who provided needed perspective at the intersection of global studies, technology, and society, while I provided a global business perspective.

The three of us launched the transdisciplinary project-based learning experience that would fulfill the senior capstone project requirement. Our student team comprised students from BME, business, and global and sustainability studies. The project, focused on developing a blood pressure monitor for use by pregnant women at risk of pre-eclampsia in Ghana, presented a threefold challenge: (1) to develop a device that not only met user needs but could also be locally manufactured and sustained over time; (2) to truly foster co-creation within the transdisciplinary team and with the network of local partners and other stakeholders; and (3) advance the project so that it met WPI's engineering accreditation requitements. Below I detail the insights from this experiment.

Forging an Integrated, Co-created Design and Prototyping Process

We began by introducing students to the integrative engineering leadership approach, including the process of empathic inquiry and co-creation that would be used throughout to drive insights and design. We invited students to develop a team charter that addressed all three spheres of ambiguity: task, relational, and process, highlighting the importance of making sure all voices were heard and the importance of teasing out different assumptions on all three ambiguity dimensions. Based on our experience, we highly recommend that this is a mandatory process, guided by faculty and completed at the very beginning of the project, ideally during an immersive multi-hour retreat. To facilitate students' awareness of differences in assumptions and introduce them to facilitation tools for forging shared understanding across disciplinary and cultural boundaries, we had each student 'mind map' how they would approach the project – outlining the different tasks and analyses needed from their disciplinary perspective – and why. Students then engaged in a facilitated discussion where they shared out their respective mind maps and then worked together to integrate their individual plans into a strategic integrated design process. We found that it was not enough for students to engage in this initial synthesis – the integrated design process needed to be further refined and reinforced (especially by faculty advisors) throughout the multiple-month long project arc. The integrated design process developed by students shown in Figure 2 became an invaluable touchstone and outcome of the capstone project.

At the same time, students collectively oriented themselves to the local context and problem space, including Ghanaian socio-cultural norms, legal and policy context, the organization of the healthcare industry, and disparities in healthcare, technology, and financial resources across different regions in Ghana.

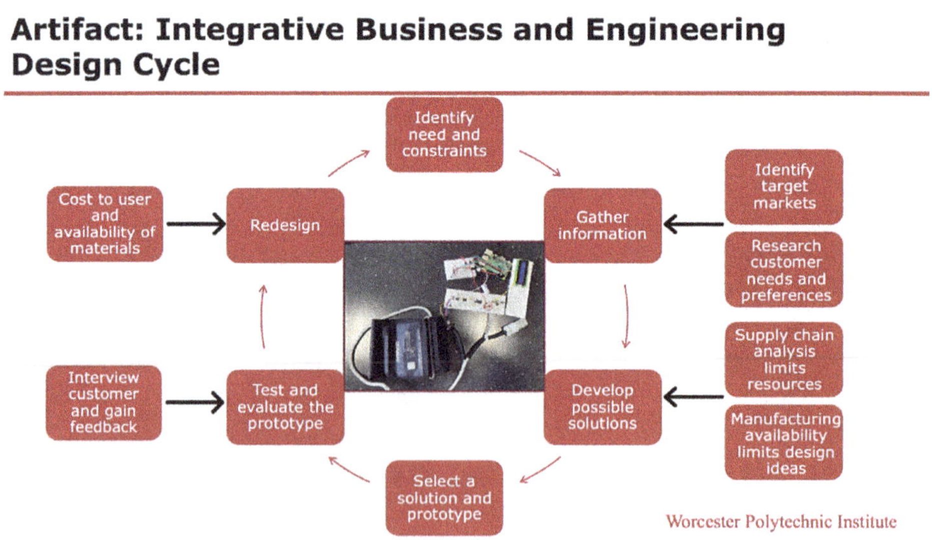

Figure 2: Integrated engineering design process.

We highly recommend that students gain this immersive local insight as early as possible within their project experience through cultural immersion and interviews with local actors rather than solely through background research. We recommend setting up a network of local partners related to the engineering problem space prior to the project so students can begin empathically interviewing and co-creating with local experts from project inception – or even before as part of essential pre-term orientation.

Developing Capacity to Elicit and Embrace Multiple, Interdependent, and Emergent Stakeholder Considerations

After initial background research, we invited students to develop interview protocols for their empathic interviews. We highly recommend having social scientists trained in conducting empathic interviews help engineering students learn how to choreograph and engage in empathic interviews and how to craft questions that engage local actors in co-creating solutions. We suggest allocating two to three weeks for protocol development, including interview role play among students and reflecting upon and refining the protocol and approach based on the role play experience. Even with guidance, students needed to iterate their initial protocols several times. We then invited students to empathically interview key Ghanaian partners to identify unmet needs, explore potential constraints and resources, and develop further individuals to interview, and engage in further reflection and protocol refinement. We also encouraged students to map the more expansive network of stakeholders, including those who might impact the design, resourcing, manufacturing, implementation, marketing and sustainability of the project. We asked students to identify what they knew and did not know about each stakeholders' perspective to drive future empathic interviews and discovery that would inform all aspects of the integrated design process.

We found that students wrestled with how to share and make sense of the volume of data they obtained from ongoing interviews and the ambiguity regarding how to move forward as they worked through multiple, interdependent stakeholder and design considerations that emerged as interviews progressed. For example, engineers were eager to scope out potential technical possibilities for storing blood pressure data and communicating it to doctors but such analysis was dependent on the choice of target user and the extent to which users had access to cell phones, internet, and medical care. As another example, interviews with e-waste brokers revealed that certain components that engineers hoped to use in their device were not available in Ghana; importation tariffs made importing the parts cost prohibitive. As a result, engineers had to be flexible about the exact specifications of their components and adjust plans as appropriate based on newly identified material sourcing constraints. This in turn could have ripple effects in their emerging design. Cost and durability design constraints also could not be fully resolved during the project arc – interviews revealed the potential for the Ghanaian government to subsidize users' purchase of blood pressure monitors if the monitors were made locally. However, it was unclear what durability and cost requirements would be needed to make this a possibility. Finally, initial interviews revealed that electrical circuit manufacturing was not prevalent in Ghana, however a later connection with a student team at a local university opened up new avenues for co-creation and production through a private/public partnership.

To help students navigate these challenges, it is imperative that faculty advisors help students think through multiple possible design scenarios and possibilities, the questions that need to be answered to winnow down options, and what best alternatives for moving forward might be if some ambiguity could not be fully resolved. We recommend engaging in these conversations weekly in the initial parts of the project – constantly updating and reorienting action based on insights gleaned from interviews or other data collection and reinforcing the interdependency among the disciplines. Importantly, we also recommend that faculty advisors be clear about when students need to move forward despite the ambiguity – and help them work through the best possible options while also recognizing that some issues will need to be resolved beyond the scope of the project.

Fostering a Community of Practice to Support Integrative Engineering Leaders

Most BME projects at WPI build upon already identified problems' spaces and insights from prior project prototypes – projects tend to be much further along in the needs-identification and prototyping process. The ambiguous and multidimensional nature of our project's needs-finding process caused considerable challenges for our transdisciplinary team as our engineers compared where they were in the design and prototyping phase with other BME senior capstone teams. Feeling like they were always 'behind' their engineering peers caused considerable stress and exacerbated existing tendencies to jump to design and prototyping phases rather than engage in integrative engineering leadership.

Going forward, we seek to develop a community of practice around transdisciplinary MQPs and integrative engineering leadership. In addition to leveraging an existing set of local stakeholders that students can engage from project inception, having a cohort of engineers engaged in similar projects will be invaluable for sharing best practices, normalizing the complex, multidimensional design process, and developing a shared understanding of expected milestones and expectations for these projects – while still meeting accreditation and graduation demands.

In the end, our students raved about this unique and powerful engineering project experience. While messy, complex, and rife with ambiguity, students felt this experience better prepared them for the real-life challenges of working across disciplines, co-creating with local users, and generating workable prototypes that could be implemented and sustained locally over time within low-resource settings.

Juan Lucena

Chapter 10 Can Community Development Projects in Engineering Education Be Both Responsible and Sustainable? Theory, Education, and Praxis

Introduction

For 75 years, institutions of higher education have been churning out economists, civil engineers, business professionals, and public and nonprofit program developers and managers to staff the organizations of international development. In the previous chapter we discuss the kinds of empathic leadership skills that could produce more empathic development professionals from those disciplines who have previously been part of the problem. Rather than further inculcate students in the belief that one – size development fits all engineering education has begun to take a cultural turn by integrating social responsibility and sustainability into the practices related to development. In this chapter, we examine this direction and its implication for sustainable outcomes.

In engineering education, there has been a proliferation of projects aimed at engineering for community development (ECD), most of these motivated by strong personal desires to help solving big and complex problems like poverty, climate change, lack of drinkable water, etc. and also by institutional needs to enhance students experiences and professional development, exposure to international education, and, in some cases, to contribute to program accreditation.

But what if these problems are so complex that they cannot be solved by engineering projects alone, and certainly not by engineers in training operating in highly constrained educational environments (e.g., 50–75 min classes, 14-week semesters, and design classes away from communities) and under limited resources of time, money, and expertise? Where and how can/should engineering educators focus their attention and resources to develop, organize, integrate, and support these projects so they can be more effective and sustainable into the future? What kinds of criteria should guide how we teach students to assess the communities they want to work with? And once they identify potential communities, what criteria should guide their behavior and the projects to serve those communities? Can these criteria be taught in the classroom and/or implemented in project spaces in our engineering programs? This chapter aims at answering these questions and providing a roadmap to organize and deliver community development projects in a responsible and sustainable way.

https://doi.org/10.1515/9783112218716-013

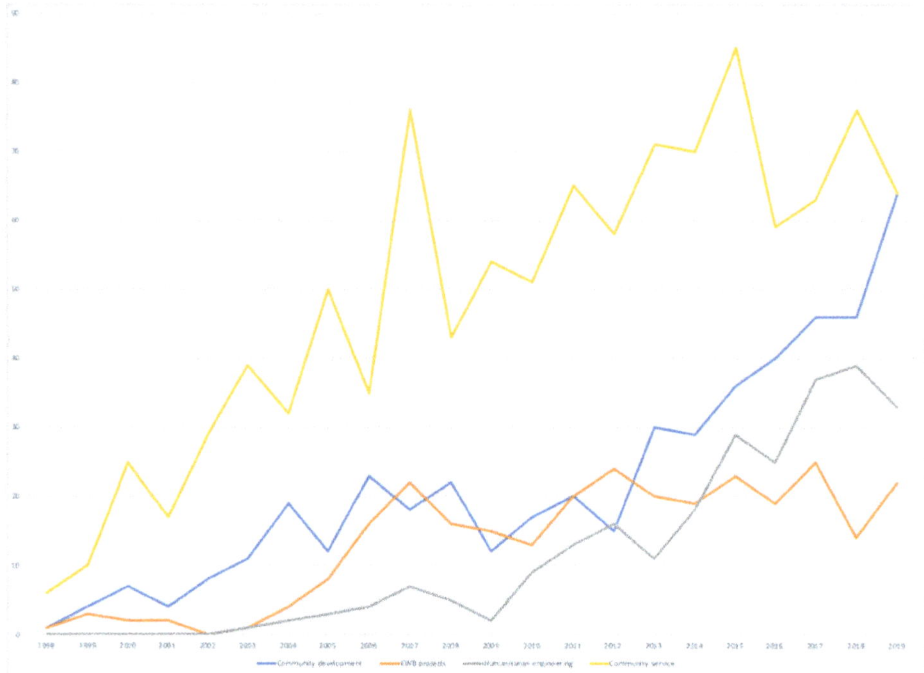

Figure 1: Growth of number of papers presented and published in ASEE conferences between 1998 and 2018 on projects related to community development, EWB, humanitarian engineering and community service (source: peer.asee.org database).

Motivation of Engineering for Community Development (ECD) Projects

Marked by the emergence of *'Engineering to Help'* programs (EWB, humanitarian engineering (HE), community development, community service, etc., ca. 2000, as shown in Figure 1), ECD projects have been motivated by faculty and students desire to help, personal and career goals, desires to study and work abroad, and desires to solve problems and to gain hands on experience on impactful work (Guzak et al., 2012, Schneider et al., 2009). Since then, some scholars have called our attention to how the focus of well-intentioned ECD projects on technological fixes and deliverables tend to leave out critical reflections of engineers' motivations to be in these projects, and of the processes required to build trust and determine communities' priorities and desires (Nieusma and Riley, 2010, Lucena et al., 2010). Unfortunately, these calls to critical reflection in the ECD space are often overshadowed by the continued emergence of milestones and challenges (e.g., UN Sustainable Development Goals and NAE Grand Challenges), ever-changing institutional priorities (e.g., the craze over innovation and entrepreneurship in engineering education), many of which are uncritically adopted as givens even when these hide significant power dimensions that put the communities we are trying to serve at a disadvantage (Does America Need More Innovators, Struckmann, 2018).

At the same time, students involved in ECD projects often lack the required sophistication to understand the complex structural forces behind large problems like poverty and climate change, and naively participate in ECD projects without normative guidelines for how to identify communities, how to behave when engaging communities, or how to develop projects that will be conducive to socially just and sustainable community development. So what are educators and their students to do? Instead of pretending that ECD projects are going to make a dent in solving the world's complex problems like poverty or climate change, just because they might be inspired by SDGs or some form of engineering grand challenges, we should develop the humility to accept that ECD projects won't solve these large structural problems and yet they might make a positive difference in communities' ability to deal with poverty or climate change. How can we teach our students this humility and ability to identify communities where they can make a positive difference?

Theory 1: Resilient Communities

Most of the structural problems affecting poor communities (homelessness, violence, climate change, hunger, etc.) cannot be solved by engineering alone. While engineers need to understand the structural conditions in which they operate (see SRE Criterion 1), we need to be careful when promising technocentric solutions to large complex problems since 75 years of international development have shown us that most of these solutions do not work (Moyo and Ferguson, 2010). Instead, when facing these complex problems, engineers can be better positioned to contribute to the resiliency of communities to adapt to the consequences of large complex problem like poverty or climate change. Resiliency can be understood as the human capacity to face and overcome adversity (e.g., the consequences of poverty), and, hopefully, come out stronger and transformed even while staying in poverty (Goldstein, 2012; N. R. Council, 2011). While the concepts of 'resilience' or 'resiliency' are now in vogue in engineering academic circles, these are seldom understood, especially in relationship to ECD projects. If we want to enhance resiliency in communities through our engineering projects, we have to be careful that our efforts do not become exclusively technocentric. A preliminary review of the literature in resiliency in ECD projects shows that most place resiliency as either a characteristic of the built environment (infrastructure) or as the attitude that engineering students must develop to succeed in a demanding curriculum (Salzman et al., 2018; Samuelson and Litzler, 2016). But how about resiliency as a latent characteristic of communities that can be both enhanced by the way engineers behave when working with communities and by the ECD projects they define with and deploy in those communities?

First, we need to recognize three essential categories of elements that need to be present in a community in order to begin building resiliency: extrinsic, intrinsic, and relational (Collaborative Resilience, Callaghan and Colton, 2008; Resiliencia. Descubriendo las propias fortalezas). Using examples from a course titled Responsible Engineers, Socio-technical Systems, Resilient Communities, funded by an NSF project aimed at teaching engineering students how to work with artisanal and small-scale gold mining (ASGM) communities in Colombia and Peru, I will outline these categories and briefly explain how to prepare students to identify these categories in order to assess whether or not a community has the necessary conditions for resiliency (see section **"Education: Curricular Integration of SRE and ESCD Criteria in Engineering Classroom"** for full description of the course).

Extrinsic: If communities have not experienced external adversity, they do not have anything to be resilient to. Hence in order to start building resiliency, communities need to have experienced external adversities such as traumas, disasters, and threats that come from large systemic of structural problems like poverty, climate change or violence from organized crime. For ASGM communities, these often include the threats from criminal activity related to ASGM (e.g., financing of violent groups through gold mining), neglect of state organizations, corruption, extorsion, systemic discrimination (of women miners, for example), chemical toxicity such as mercury poisoning, the presence of multinational corporations that compete for mineral resources, geological hazards, among many others (Pieth, 2019; Veiga and Marshall, 2019; Hilson, 2002). Students in my class learned about these external adversities as follows. First, they read and understood the linkages between poverty and ASGM (Siegel and Veiga, 2010) and how the conditions of poverty and illegality in ASGM are often perpetuated by systemic problems in the formalization of the ASGM sector (Veiga and Marshall, 2019). Second, following community connections that faculty developed for more than a year, students connected with ASGM community members (via WhatsUp or Skype) to see how they experienced these adversities. This understanding of and connecting with ASGM communities proved very effective during the subsequent summer field trip (see section **"Praxis 1: Problem Identification with ASGM Communities"**).

Intrinsic: For resiliency to occur, there has to be a history of positive adaptations intrinsic to a community that has tried to overcome adversity. In ASGM communities, these have included mercury detoxification programs (Ruiz Córdoba et al., 2016), territorial local planning that keeps ASGM away from watersheds (Restrepo et al., 2013), the desire of groups that have been discriminated to organize to gain voice and power (Smits, 2019), etc. For these types of positive adaptations to occur, communities must exhibit the following characteristics: Collective self-esteem (love who you are as a community); Cultural identity (know who you are as a community); social humor (be able to laugh at who you are); and state honesty (transparency between local/state government and communities) (Resiliencia. Descubriendo las propias fortalezas). Students in my class learned about these positive adaptations by researching the above characteristics through local newspapers and radio stations where collective self-esteem and humor can be identified, and local-government websites that exhibit different degrees of state honesty. For our particular ASGM community in Colombia, they learned that there are high levels of collective self-esteem and humor, significant transparency of the local government but that state and national government processes related to ASGM are obscure at best, hence presenting a significant obstacle for resiliency of communities. To overcome the hurdles of state bureaucracy and lack of transparency, many ASGM communities have to resort to informality and illegality (Veiga and Marshall, 2019) and students have to define problems and propose solutions within this reality.

Relational: This element refers to social trust as the main element that needs to be present between a community and outside experts (like engineers) to build resiliency. For engineers working with ASGM communities, for example, in order to gain social trust they need to show: (a) competence, not only in an engineering domain but also in building and managing the social relations that make projects possible; (b) caring, not only about the technologies but also about the people impact them by them; (c) predictability, showing that you are going accompany and be accountable to ECD projects from start to finish and thereafter; and (d) commitment to diversity of perspectives and interpretations (e.g., various representations of the problem and possible solutions) (Collaborative Resilience). Students in my class learned about these elements through group exercises where they were challenged to develop interactive activities with communities where students could present their competence, caring, willingness to accompany communities for the duration of the project, and commitment to include many voices during the different stages of a project. These activities were piloted in class and will be validated in direct contact with communities during summer field work.

While engineers have little, if any, control on the intrinsic and extrinsic elements of resiliency, they need to able to identify them and consider them as contextual constraints in their problem definition and designs. Certainly, engineering students can have significant influence over the relational dimension by cultivating social trust by exhibiting competence, care, predictability, and commitment to diversity. But in doing so, what should the guiding principles of students' behavior with communities be?

Theory 2: How Should Engineers Behave with Communities?

In a different writing, we have shown how engineering ethics benefit the relationships that engineers have with corporate employers while not serving as appropriate guides in their relationship with communities (Smith and Lucena 2021). To overcome the limitations of engineering ethics (the codes and the forms in which they are taught), we developed a set of criteria for socially responsible engineering (SRE), which I highlight here with examples of how students began developing these in their role as engineers who will be working with ASGM communities:

Criterion 1: Understanding structural conditions and power differentials among specific stakeholders of an engineering project. In our course, students researched and understood how mining law and codes benefit large scale mining companies more than ASGM miners, or how among ASGM miners, men have more power and benefit more from mining than women miners (Normatividad – Minería en Colombia). Students were able to validate these research findings through in-class teleconference exchanges with engineers from large mining companies and women miners.

Criterion 2: Contextually listening to all stakeholders, especially those who are marginalized, to grasp their needs, desires and fears surrounding a specific project. In our course, students listened how to the histories of miners, paying particular attention to how they got there, what their struggles are, and to the stories of disempowered groups like women and indigenous populations who have been unequally affected by mining (Colonia, 2017; libreriadelaU.com). To get this knowledge, our students interacted in person with an anthropologist who has studied and written about ASGM in Colombia (Ferry and Ferry, 2017) and a lawyer-activist who fights for the rights of ASGM communities, and virtually with a number of other actors invested in the well-being of disempowered groups in ASGM (Leadership, Organizing and Action – Women & Subsistence Mining in Medellín, Colombia | MIT D-Lab).

Criterion 3: Collaboratively identifying opportunities and limitations of creating shared social, environmental and economic value for all stakeholders, especially those who are marginalized. We are teaching our student cohorts how to do this by showing them how others have organized and delivered workshops (Leadership, Organizing and Action – Women & Subsistence Mining in Medellín, Colombia | MIT D-Lab) to bring different stakeholders together to identify opportunities that contribute to social (e.g., creating associative corporations), environmental (e.g., reduce use of mercury), and economic (e.g., increase income diversity) values. Then students can emulate these collaborative practices in their own ASGM-related projects (Smits, 2019).

Criterion 4: Adapting engineering decision-making to promote those shared values, acknowledging situations in which this is not possible and when engineering projects should not move forward. In our course, we taught students how if one dimension of ASGM stops being profitable, safe and/or environmentally sound (e.g., dispersed gold processing plants that might be polluting different watersheds), engineers can engage communities in proposing and jointly create alternatives (e.g., a centralized processing plant that benefits from economies of scale for all and environmental supervision). Unfortunately, in the ASGM areas where we work, these alternatives have been developed solely by engineers without the input from communities. Our students are learning about the perils of not engaging communities.

Criterion 5: Collaboratively assessing activities and outcomes with those stakeholders. In ASGM contexts, this could be done by teaching students how, for example, if the process to decide, design and build a centralized processing plant is the result of a joint effort between engineers, ASGM communities, and local government, then a collaborative process for continuous feedback and improvement should be put in place to assess the costs and benefits of the plant. In our course, we played out a hypothetical scenario where I invited students to imagine that the centralized processing plant described above had been designed with community input and then invite them to "describe how you will adapt engineering decision-making to promote those shared values, acknowledging situations in which this is not possible to do so and engineering projects should not move forward. Please be specific about engineering decision-making. Remember that engineers build systems, processes, infrastructures, technologies, etc. that legislate, even after elected officials end their terms".

How do these SRE criteria map to social trust? Without doing an exhaustive analysis of how each criterion contributes significantly to building social trust, one can see that, for example, contextually listening to all stakeholders (criterion 2) contributes to competence, as engineers would understand how a proposed project would affect stakeholders differently (and vice versa); caring, as stakeholders would see how engineers care about their histories, struggles and hopes in relationship with the proposed project; predictability, as engineers would come to understand the importance of 'being there' or 'coming back' for different stakeholders; and commitment to diversity, as engineers would understand how the jointly defined problem and proposed solutions would mean different things, have different interpretations, and elicit different predictions to different stakeholders (see Table 1).

But even if we can develop social trust between resilient communities and socially responsible engineers there is no guarantee that jointly defined projects will actually contribute to community development in a sustainable and socially just way. What kind of criteria should guide engineering projects aimed at sustainable community development?

Table 1: How each SRE criterion contributes to different elements of social trust.

Elements of social trust SRE criteria	Competence	Caring	Predictability	Commitment to diversity
1. Understanding structural conditions and power differentials	✓	✓		✓
2. Contextually listening to all stakeholders	✓	✓	✓	✓
3. Collab. identifying opportunities and limitations of creating shared SEE value	✓	✓	✓	✓
4. Adapting engineering decision-making to promote those shared values	✓	✓	✓	
5. Collaboratively assessing activities and outcomes with those stakeholders	✓	✓	✓	✓

Note that the check marks can be replaced with actual numerical ratings based on survey data that students can gather by deploying questionnaires in communities. Our new cohort of engineering students learn these survey techniques in a community-based research course.

Theory 3: Guiding Criteria for Engineering Projects with Communities

According to rural sociologists Bridger and Luloff (1999), sustainable development has not been achieved since its adoption in the early 1990s because its well-intentioned goals and practices have been appropriated by the technocrats of international development who live and work far removed from the daily struggles of most people on the planet (for a full critique of sustainable development discourse, see Bridger and Luloff (1999). For a critique of the planning mentality behind sustainable development, see Easterly (2006)). Hence they challenged us to conceptualize, develop and practice sustainable development in the localities most affected by macroscale phenomena as climate change and social inequality: communities. They write:

> [b]y shifting the focus on sustainability to the local level, changes are seen and felt in a much more immediate manner. Besides, discussions of a 'sustainable society' or a 'sustainable world' are meaningless to most people since they require levels of abstraction that are not relevant in daily life. The locality, by contrast, is the level of social organization where the consequences of environmental degradation are most keenly felt and where successful intervention is most noticeable . . . sustainable community development may ultimately be the most effective means of demonstrating the possibility that sustainability can be achieved on a broader scale, precisely because it places the concept of sustainability in a context within which it may be validated as a process. By moving to the local level, the odds of generating concrete examples of sustainable development are increased. As these successes become a tangible aspect of daily life, the concept

> of sustainability will acquire the widespread legitimacy and acceptance that has thus far proved elusive. (Bridger and Luloff, 1999, p. 380)

Bridger and Luloff proposed five criteria for sustainable community development that I have further developed with my students, making them appropriate for and operational in engineering projects, which we now call Engineering for Sustainable Development Community (ESCD) criteria. While these criteria are by no means all encompassing, we have found that they are comprehensive and effective to get novice engineers to focus on dimensions that are truly important in the communities where we work. In our ASGM projects, our students are encouraged to define and develop projects that incorporate all of these ESCD criteria and identify conflicts among them. The five ESCD criteria, with specific examples related to ASGM communities, are:

Local economic diversity (LED): This criterion includes: (1) the creation of new local products/markets, which for ASGM communities could include the making of bricks, adobe, and other construction materials from remediated mine tailings; (2) production based on local resources, processes, relationships such as buying the equipment to remediate the tailings and make the bricks in local businesses so revenues are reinvested in the local economy; and (3) capacity building and diverse local job creation such as the formation of local masons and brick makers that can benefit from the new locally available materials (Roy et al., 2007).

Political self-reliance of communities: This criterion includes: (1) autonomy in decision making such as the kind enhanced by associative organizations where ASGM communities come to their own decisions and voice them in front of more powerful actors like government agencies and multinational mining corporations; (2) reducing dependency from external expertise and knowledge as it happens when ASGM miners engage our students on a leveled playing field to co-define problems and solutions together. In our work in ASGM communities, we encourage our students to link their projects to associative corporations (e.g., like the association of artisanal miners) so these can vet, support, help manage, and ultimately decide how and if the project should continue into the future (Leadership, Organizing and Action – Women & Subsistence Mining in Medellín, Colombia | MIT D-Lab).

Reduction/reuse of energy and materials in locality: This criterion includes: (1) conceiving projects through the 'cradle-to-cradle' approach (McDonough and Braungart, 2010); (2) improve energy efficiency and curb energy consumption in the design of the project sub-systems; and (3) include energy and materials storage capability to use excess energy and materials later. In our ASGM projects, we challenge students to incorporate organizations committed to the elements of this criterion such as securing energy from renewable sources to minimize dependency from the grid and/or involve organizations of women miners ('chatarreras') who can profit from recycling waste material in mines and processing plants (Laplonge, 2017).

Enhance biodiversity and careful stewardship of natural resources in locality: This criterion includes ensuring that projects and their by-products (1) will not negatively impact biodiversity or natural resources such as forests and rivers; and (2) how they can actually enhance the quality of natural resources such as when mine tailings are cleaned from heavy metals, removed from the environment, and used as construction materials. In our ASGM projects, we invite students to learn about the importance of biodiversity and natural resources directly from community members by listening to local stories about the meaning of rivers, mountains, forests, etc.

Social justice in locality: For the community impacted by a project, this criterion includes: (1) increasing opportunities (jobs, health, education, etc.) and resources (clean water and air, construction materials, income, arable land, etc.); (2) decreasing risks and harms (e.g., reducing heavy metals in the environment or improving ergonomics of mining tools to prevent injuries); and (3) enhancing human capabilities such as bodily health, bodily integrity, affiliation, and so on (Nussbaum, 2011). (For an extensive analysis of how to incorporate this criterion throughout the engineering curriculum, see Leydens and Lucena (2017).)

Education: Curricular Integration of SRE and ESCD Criteria in Engineering Classroom

The motivation to bring ASGM into engineering education started with a NSF-partnerships in international research and education (PIRE) grant titled "Responsible Mining, Resilient Communities" (RMRC), an inter-institutional, interdisciplinary, multi-country project aimed at researching, co-designing, implementing and evaluating sustainable ASGM technologies and practices with ASGM miners and affected communities in Colombia and Peru. Using the SRE, community resiliency, and ESCD criteria as organizing frameworks, I developed a course titled "Responsible Engineers, Socio-technical Systems, Resilient Communities" and its description and learning objectives reads as follows:

> Through the lens of artisanal and small-scale gold mining (ASGM) viewed as a socio-technical system, this course explores the relationship between responsible engineering and the development and maintenance of resiliency in communities that historically have been ignored or marginalized by engineers and the organizations that employ them. Through in-depth readings, class discussions and projects, students will 1) study and analyze different forms of responsibility in engineering and resiliency in complex communities like ASGM communities; 2) critically explore strengths and limitations of dominant methods in engineering problem solving, design, and research for working with these communities; 3) develop understandings of effective forms of responsible engineering to work with communities, especially through the integration of social science concepts and methods in order to understand ASGM as a socio-technical system; and 4) research, develop, evaluate and present projects on how responsible engineering can lead to resilient ASGM communities.

In Spring 2019, the course was taken by the undergraduate and graduate students selected to work in the PIRE-RMRC grant. As stated above, one of the main goals was to research, scope, define and develop engineering projects that can contribute to resiliency in ASGM communities. This goal was to be achieved, first, by developing students' understating of ASGM as a socio-technical system. To do this, we applied existing conceptual frameworks from science and technology studies (STS), mostly developed in the US and Western Europe, to understand socio-technical systems (Bijker et al., 1987; Callon, 1987) and came to the realization that these conceptual tools are very limited when we want to influence systems for purposes of equality, resiliency and social justice. We discovered that STS literature on socio-technical systems (the socio-technical analysis framework) developed in the Global South (Thomas and Santos, 2016) was more appropriate to understand ASGM in Latin American countries and to conceptualize the relationships between responsible engineers, resilient communities, and sustainable projects with the goal of improving the well-being of ASGM communities in Colombia.

Alongside this conceptual exploration of ASGM, students interacted with ASGM miners, community members, and engineers working in ASGM settings, via teleconferencing technology (Skype and others), and texting (WhatsApp) in order to validate their learning of ASGM as a socio-technical system, and identify relevant problems to ASGM communities from different stakeholder perspectives. Students synthesized their conceptual exploration and stakeholder interactions in the final paper for the class where I invited them "to revisit, apply and connect two cornerstones of this course: The criteria for socially responsible engineering (SRE) and the criteria for engineering for sustainable community development (ESCD) to your chosen project. SRE is about your attitude, knowledge and skills as an engineer working with communities. ESCD is about projects that might have a better chance to improve the living and working conditions of communities." Students successfully completed this synthesis as reflected by the average grade for the course in this final paper (90%). This conceptual exploration, validation and preliminary problem identification served as preambles for full project identification during summer field work.

Praxis 1: Problem Identification with ASGM Communities

During the summer 2019, we took the group of PIRE-RMRC students to ASGM communities in Colombia: five undergraduates and three graduate students from Mines, five undergraduates from US Air Force Academy, one graduate student from University of Colorado-Boulder, and one from University of Texas-Arlington. Prior to traveling to the ASGM communities located in an Andean mountain region of Antioquia, Colombia's main gold mining region, students spent one week in Medellin (Antioquia's capital) learning about the history and political economy of gold mining, directly from local faculty and students, mining engineers, mine owners, and local officials. Two PIRE graduate students ran a mini-workshop where they challenged undergraduate groups from the US and Colombian universities to conceptualize the supply chain of gold mining and then to compare their different understandings of the supply chain given their different national origins and institutional locations.

After this one-week introduction of the ASGM context in the capital city, faculty and students traveled to Andes, Antioquia, a coffee and gold mining municipality aiming to become Colombia's first county free of mercury from gold mining. During this week, US and Colombian students partnered to:

- interact with miners, engineers, communities, local officials;
- travel to mines deep in the mountains to get a sense of the hardships involved in getting to the mines sites;
- visit processing plants where they took water and soil samples and learned about processes by talking to workers and plant owners;
- attend community meetings with women mining leaders in order to identify problems of importance to local miners but viewed from different perspectives;
- debrief their experiences at the end of the day and begin creating a map of the region where they could locate actors, communities, mining sites, natural resources (watershed, forests, etc.), sites of economic activity, etc.; and
- contextualize their problem identification with respect to human activity and the natural environment.

Praxis 2: Generating Problem Solutions and Potential Designs in the Classroom

After 10 days of problem identification, faculty and students returned to the United States with projects defined in the following problem areas: mine air monitoring, mine tailings remediation, miners' safety and health, and child education in mining communities. These identified problems found curricular space inside two design courses in Mines' Humanitarian Engineering (HE) program where PIRE and non-PIRE students together work on generating problem solutions, conceptual design, and prototyping.

EDNS 401: Projects for People. This course was developed for the HE program after realizing that traditional senior design courses are historically, organizationally and conceptually developed with assumptions that the final design is for commercial purposes for a for-profit client and not necessarily to address social change in vulnerable communities. As I have written elsewhere, engineering design education for industry, specially its assumptions, design constraints and purposes, presents significant problems for design for communities (Lucena et al., 2010). In this course, students learn to "work with innovative organizations dedicated to community development to solve major engineering challenges [and generate] engineering solutions to real problems affecting real people in areas central to their lives" (Engineering, Design, and Society < Colorado School of Mines).

A PIRE RMRC-project faculty member taught the class in the Fall semester following the summer field trip dedicated to problem definition with ASGM communities. Through more in-depth stakeholder engagement, particularly by encouraging students to regularly check their design assumptions with ASGM community members and students in Colombia, students in this class developed plausible solutions in the three problem areas as follows:

- For mine air monitoring, students developed a prototype of a low-cost air monitor that miners can attach to their helmets or shirts.
- For mine tailings remediation, students developed a conceptual design to use remediated tailings as raw material to make construction bricks.
- For miner's safety and health, students identified a number of ergonomic backpack designs to minimize back injuries when carrying ore out of mine shafts.

EDNS 491–92 ECD Capstone Design Studio: To continue addressing the problems associated with design for industry mentioned above, we developed a design studio environment inside of our university's Capstone Design course that allows us to bring in multiple ECD projects at the same time, requiring different time frames for their completion, and different skill sets required from students. The students in this design studio are supported by faculty with backgrounds in anthropology, chemistry, engineering, and STS. The two main faculty members in charge of the ECD studio are developing course content and activities to achieve the following student learning outcomes (mapped in relationship to ABET criteria at the time):

1. develop a project plan (including problem definition, design, and project completion) that can be accomplished during the allotted time period;
2. define problem, undertake research, and apply appropriate engineering knowledge to solve a design challenge in partnership with a community;
3. gather information and communicate with community members and other stakeholders about issues related to problem definition, engineering design, analysis, decision making, and solution professionally in writing using methods, including design documentation packages, synthesizing feedback and putting it into action;
4. gather information and communicate with community members and other stakeholders verbally using methods, including formal design review events, synthesizing feedback and putting it into action;
5. identify and analyze the ethical, environmental, societal, and/or economic impacts that your engineering interventions can have for community partners and other stakeholders, and incorporate these findings into problem definition, engineering design, analysis, decision making, and solution;

6. function effectively on a team whose members together provide leadership, create a collaborative and inclusive environment, establish goals, plan tasks, and meet objectives; and
7. evaluate your success as a designer by reflecting on your experience and proposing strategies for future self-improvement.

For the first semester (EDNS 491) of this two-semester Capstone sequence, the projects came directly from three sources of problem identification: a) in-depth problem identification with communities during summer field work; b) EDNS 401 Projects for People course; and c) new interactions and problem identification at a distance with local communities and in-country stakeholders (see Figure 2). During this first semester, students continued to further define the problems with community members and other stakeholders in order to have a well-defined problem and conceptual design by the end of the semester.

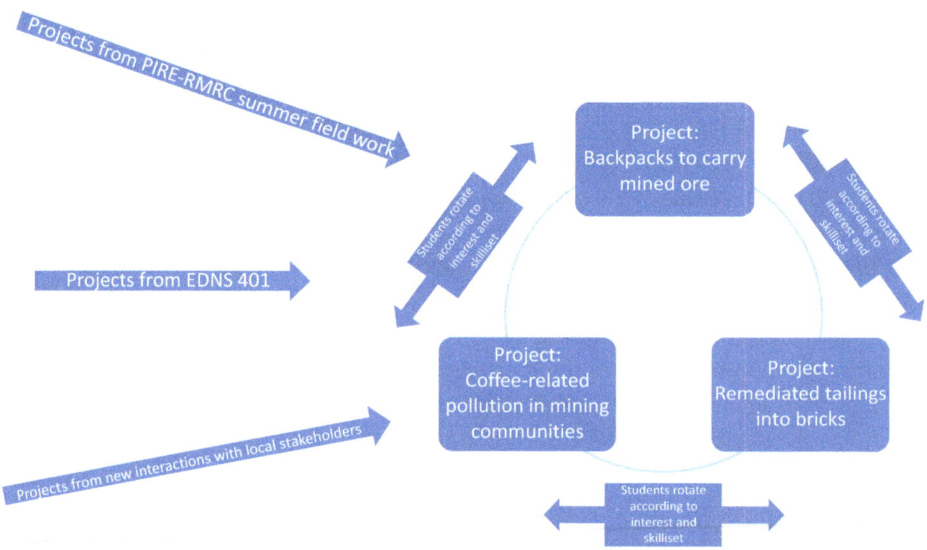

Figure 2: ECD Design Studio model.

In the second semester (EDNS 492), students continued working toward the achievement of the learning objectives above and developed design prototypes that the new cohort of PIRE-RMRC students will take to Colombia to be validated with ASGM community members (see below). This ECD design studio model also allowed for projects to flow back and forth between students in US and ASGM communities abroad, and ensured that their continuity, refinement, scaling (up or down) and the emerge of new projects have a stable curricular space.

Conclusion

As the number of engineering schools with initiatives (courses, projects, and programs) related to community engagement continue to grow (see Figure 1), engineering educators need to embrace a more humble attitude towards the problems that we seek to solve. Instead of pretending that big problems like poverty or climate change have engineering solutions, we have to acknowledge that some of these problems might have no solution at all and instead focus our energies on improving the ability of communities to respond to these problems (resiliency). As the complex problem area of ASGM illustrated above shows, there might not be a solution to the poverty that serves as the reason for many people to become ASGM miners in the first place, or the poverty that ASGM continues to reproduce (Siegel and Veiga, 2010). Hence, our job as engineering educators should be to educate our students to understand the limits of engineering solutions and perhaps direct their efforts to build resiliency in communities, using normative criteria such as SRE and ESCD criteria to guide their behavior and projects.

Furthermore, as community engagement initiatives grow, whether in the form of EWB student chapters (Jaeger and LaRochelle, 2009) and projects or minors and certificates, engineering educators and administrators should create the appropriate institutional spaces where the theoretical, educational and practical (praxis) dimensions of these initiatives can coexist and interact with each other. This chapter illustrated one model in which these dimensions interact yet there must be others.

Throughout the development of our HE program, its courses and projects, and our multiple interactions with engineering educators, students, and communities around the world, it has become increasingly clear that community engagement needs theoretical and normative frameworks to guide the action of our students as they engage communities (such as the SRE criteria above) and the expected outcomes of the their projects (such as the ESCD criteria above).

Mimi Sheller

Chapter 11
Power, Participation, and Co-design in Development Engineering and Disaster Recovery: Lessons from Haiti

Introduction

Stakeholder participation and citizen involvement have been widely embraced as ideal policies for a range of development engineering situations including water management, sustainable development, and green infrastructure.[1] Historically, and especially in developing world settings, infrastructure designed and implemented from the top-down has often failed and/or led to the marginalization and demoralization of the intended beneficiaries.[2] Recent studies demonstrate that stakeholder-based participatory processes produce better outcomes than classical top-down authoritative planning processes.[3] The participation of local stakeholders in the conceptualization and design of infrastructure projects acknowledges the 'stake' that the intended beneficiaries have in the project's outcome and, research suggests, increases the likelihood that these investments will directly address local needs and capacities.[4] In the field of post-disaster reconstruction too it is argued that the 'conventional top-down approach, by governments and large relief agencies coming in with large projects' is being rejected in favor of 'more involvement of local people' and community-driven projects.[5] In the previous two chapters, we have examined empathic leadership and called for teaching humility over hubris. In this chapter, we bring those together in the context of stakeholder engagement.

Likewise, much current research on climate adaptation and resilience also calls for not only new convergent interdisciplinary research approaches but also the co-

---

1 Barreteau et al., "A Framework for Clarifying 'Participation'"; Creighton, *The Public Participation Handbook*; Daniell, *Co-Engineering and Participatory Water Management*.

2 For example, the failure of soil and water conservation planning in the Chemoga watershed in rural Ethiopia was found to be due in large part to unwilling and disinterested participation of local farmers in an externally designed plan to construct structures widely perceived as ineffective on the ground (Bewket and Sterk, "Farmers' Participation in Soil and Water Conservation Activities"). See also Scott, *Seeing Like a State*.

3 Lynam et al., "A Review of Tools"; Reed, "Stakeholder Participation for Environmental Management"; Voinov and Brown, "Lessons for Successful Participatory Watershed Modeling"; Voinov and Bousquet, "Modeling with Stakeholders."

4 Daniell, *Co-Engineering and Participatory Water Management*.

5 Lawther, "Community Involvement in Post Disaster re-construction".

https://doi.org/10.1515/9783112218716-014

creation with communities of effective systems to preempt, prepare for, track, and respond to climate-induced hazards and vulnerabilities, and to recover from their devastating effects. Participatory approaches that engage local stakeholders, sometimes through citizen science, are increasingly viewed as necessary elements of transnational development engineering projects. Community-based participation in development engineering projects is argued to provide better and more comprehensive solutions to complex problems that cut across coupled human physical systems. It can also help to build in local capacity, and ultimately can shore up the sustainability of the entire infrastructure system and thereby reduce social vulnerability when the next disaster comes along.[6] Equitable community participation is especially crucial in the period of transition from emergency relief into post-disaster reconstruction when local community knowledge, skills, and capacity can be incorporated into rebuilding processes in order to support long-term sustainability goals that include social equity, the reduction of social vulnerability and hence future risk reduction.[7]

Ideally, a collaborative approach between stakeholders, communities, government agencies, engineers and scientists can ensure more effective, equitable and robust infrastructural projects by producing tools, processes and governance approaches that not only are directly beneficial to all stakeholders, but also are produced together and validated through dialogical processes of knowledge creation. Crucially, though, this requires that all participants acknowledge the social and political dynamics at play, the unevenness of citizenship and expertise, and the differential power of each participant to make choices or decisions, and to have their voices heard.[8] Participation is important but needs to be done in a way where we pay attention to power and inequities in who can speak, whose voice is heard, and who can make decisions.

This is true both of general development projects and of post-disaster reconstruction projects. Participatory engineering has become an especially crucial aspect of the imperative to 'build back better' after major natural disasters, such as the 2010 earthquake in Haiti where this slogan was adopted by international funders, but it can take many different forms with varied results.[9] My own research on reconstruction in post-earthquake Haiti informs the arguments in this chapter.[10] Studies of cases such as post-tsunami Aceh, Indonesia and Sri Lanka, or floods, landslides, and earthquakes

---

6 Sheller et al., "Participatory Engineering for Recovery in Post-Earthquake Haiti."

7 Leon et al., "Capacity Building Lessons;" and see Katz, *The Big Truck That Went By*.

8 Furlong, "The Dialectics of Equity"; McFarlane et al. "Informal Urban Sanitation"; Swyngedouw, *Social Power and the Urbanization of Water*.

9 Kennedy et al., "The Meaning of Build Back Better".

10 Galada et al., "Attitudes Toward Post-earthquake Water and Sanitation Management"; Galada et al. "Assessing Preferences Regarding Centralized and Decentralized Water Infrastructure"; Sheller et al., "Gender, Disaster and Resilience: Assessing Women's Water and Sanitation Needs"; Sheller et al., "Participatory Engineering for Recovery in Post-Earthquake Haiti."

in other small island developing states have emphasized the importance of integration of local and indigenous knowledge with scientific knowledge as a crucial basis for disaster risk reduction, especially in the face of increased place-based vulnerability due to climate change and historical processes of marginalization.[11]

Social science research is necessary in collaborative development engineering, I argue, because social scientists have expertise in assessing social inequities, in developing intercultural competencies, in recognizing power imbalances, and in bridging research and action in iterative and recursive ways. More broadly, participatory development engineering informed by social science approaches can help to ensure procedural and distributive justice and fair processes of inclusion in decision-making by mitigating uneven degrees of power. Critical participatory approaches contend that researchers, engineers, and policymakers must meaningfully engage stakeholders in every aspect of this process, which calls for new approaches to how community stakeholders are involved in engineering projects.[12] Hendricks et al. (2022), for example, call for the use of community science and public participation to "empower residents in using community-driven research, analytics, organizing, and mobilization to advocate on their own behalf and to implement community-level programs that level the landscape before, during, and after flood disasters".[13]

In disaster recovery relating to heritage preservation, Gibson et al. (2018) further argue that "The values, doctrines, and methodologies of orthodox heritage practice do not consider marginalization, segregation, and exploitation of traditionally disenfranchised groups," and that there is a need for participatory methods that acknowledge this and facilitate broader inclusion in engagement, empowerment, and decision-making.[14] Black feminist planning approaches further criticize technical measures of vulnerability that do not take into account structural racism and intersectional inequalities. Fayola Jacobs highlights first that "systemic forms of racism, sexism and classism severely impact communities' abilities to prepare for and recover from disaster";[15] and second that radical planning therefore requires the use of alternative methods of inclusion that acknowledge that "lived experience is a way of knowing and knowledge is created and validated through dialogue".[16] These are all valuable lessons for development engineering more broadly.

In this chapter, I will first review several recent approaches that have emerged for better incorporating equitable participation and inclusive approaches into devel-

11 Mercer et al., "Reflections on Use of Participatory Research"; Mercer et al., "Integrating Indigenous and Scientific Knowledge Bases"; Leon et al., "Capacity Building Lessons"; Kelman et al., "Combining Different Knowledges."

12 Arnstein, *A ladder of Citizen Participation.*

13 Hendricks et al. "Moving Up the Ladder in Rising Waters", p. 2.

14 Gibson et al. "From Engagement to Empowerment", p. 596.

15 Jacobs, "Black feminism and radical planning", p. 26.

16 Ibid, p. 29.

opment engineering and post-disaster reconstruction frameworks. Specific frameworks and methodologies to be discussed include the social amplification of risk framework (SARF), the use of vulnerability, consequences, and adaptation planning scenarios (VCAPS), resilience governance analysis (RGA), participatory action research (PAR), and citizen science. In the second part, I will draw on lessons learned in post-earthquake Haiti to assess the ongoing tensions within – as well as the promise of – building more inclusive transdisciplinary approaches to development engineering, disaster recovery, and climate adaptation. I argue that practitioners must be cognizant of the uneven power relations and structural inequalities that may hinder effective participatory processes and must actively reflect on ways to acknowledge and overcome such barriers.

Emerging Models for Community-Based Development Engineering

There are several models that have been used for community-based development engineering, disaster recovery, climate risk reduction, and climate adaptation that seek to incorporate social science understandings of structural inequalities and power relationships. Human-centered co-design is one problem-solving approach that starts with the people impacted the most by an existing infrastructural context. Often this can be informed by a social-ecological-technological systems (SETS) framework.[17] The SETS framework is a way to center the complex relationships between human systems, ecological systems, and technological systems, and has been used to inform the assessment of vulnerabilities, constructing knowledge-networks, and co-designing interdependent systems. In this section, I will give an overview of some of the frameworks and methodologies that seek to incorporate community-based participants into engineering activities, infrastructure planning, and decision making.

One human-centered approach is the SARF which is used to analyze the process of transformation in the face of risks. SARF highlights the ways that actors in risk arenas can both amplify and attenuate perceptions of risk, leading to social responses (individual and group behaviors, organizational/institutional actions), that can affect the character and magnitude of consequences from those risks by amplifying or attenuating them. Application of the SARF can add to researchers' understanding of the role of organizations – for example, municipal governments, NGOs – and of individual actors in generating new problems and amplifying (or attenuating) existing ones.

17 Iwaniec et al., "A framework for Resilient Urban Futures"; Chang et al., "Assessment of urban flood vulnerability using the social-ecological-technological systems framework."; Markolf et al., "Interdependent Infrastructure as Linked Social, Ecological, and Technological Systems (SETS)."

In other words, impacts can ripple outward such that new people and groups are impacted, and also because their actions (responses) can generate new risks and impacts.[18] It is important for development engineers to understand how the projects they undertake might amplify risks, and spin out in unintended ways, or conversely, could help to attenuate existing risks and help make communities more resilient.

Nevertheless, this approach is still crafted by planners and policy makers, and couched in the language of technical expertise. This leaves a tension between power relations and the decision-making process because it does not address uneven power relations and epistemic differences as an essential element. This it is just one tool in the toolbox, but not the only one needed. Practitioners of critical development engineering and disaster response argue that multi-directional learning among community stakeholders and researchers, scientists, or engineers, should be guided by a continuity of partnerships and multi-directional learning that can engender not only greater trust and meaningful engagement, but also more resilient outcomes that reduce community vulnerability in the future. These kinds of reflexive and socially-embedded participatory projects can co-create a dynamic 'knowledge-action network' that can help develop effective and scalable processes for improved collaboration between researchers, residents, planners, responders, and government agencies to transform hazard anticipation, disaster response, community climate adaptation, and community resilience not just for recovery in the short term but for better results in the medium to long term.

Another planning model that could assist in development engineering projects is the VCAPS process for eliciting and organizing information about the effects of hazard events and identifying opportunities to intervene to mitigate impacts.[19] VCAPS is a joint project of the Social and Environmental Research Institute and the Carolinas Integrated Sciences and Assessments Programs. In 2011, VCAPS was initially tested and refined through its application in two South Carolina coastal communities. Since then, it has been applied in additional communities in Alabama, Colorado, Hawaii, Maine, Massachusetts, North Carolina, South Carolina, and Puerto Rico with the assistance of research, outreach and extension partners. In addition, training workshops have been conducted to show local planners and extension agents the benefits of using the scenario-building approach to identify climate change adaptation strategies. Proponents argue that VCAPS can support local decision-making and strengthen adaptive capacity that seeks just, equitable, and integrative solutions. To be effective it seeks to address issues of equity, power, and influence and to negotiate shared principles among participants. The VCAPS approach can help people identify and articulate gaps

---

18 Tuler et al., "Learning through Participatory Modeling"; Tuler et al., "Assessment of adaptation, policy, and capacity building outcomes."

19 See http://www.vcapsforplanning.org/ and work by my WPI colleague Seth Tuler, Tuler et al., "Learning through Participatory Modeling"; Tuler et al., "Assessment of adaptation, policy, and capacity building outcomes."

in information and additional research questions. It also can help reveal the impacts that could (or do) emerge from different actions taken to mitigate risks and impacts, which may shift negative impacts or burdens onto other places (e.g., downstream communities) or other people. This is a process that can help both elicit and document these recursive processes.

Scenario building is another way to develop adaptation and transformative scenarios (aspirational) and the pathways that define who, what, and when decisions are needed today to transition towards those futures. RGA is a model that is currently being used for such processes to identify a wider scale of institutional governance.[20] It begins with surveys and interviews within particular communities to identify the following elements:

(1) Risk perceptions and problem framings – problems and solutions to address extreme events in cities, including flooding (coastal and urban), heat, and drought, and how resilience is conceptualized;
(2) Network mapping – the relationships between actors collaborating in climate resilience initiatives and producing knowledge related to climate change in the region;
(3) Knowledge systems – the data/information/knowledge that stakeholders use to learn about and evaluate risks and uncertainties related to climate change;
(4) Visions and priorities – actor goals and existing strategies/actions;
(5) Capacities, in terms of the existing adaptive and anticipatory capacities necessary to plan and prepare for the future; and,
(6) Organization profiles, including: type, roles and responsibilities, areas of expertise, climate resilience activities, geographic scope and level of action.

Then back-casting is used to formulate specific adaptive capacities (short-term, risk mitigating responses), but also general adaptive capacities (long-term, systemic strategies to transform development and institutional structures, such as investing in equity, health, education, housing, etc. and more radical solutions, like managed retreat, etc.). Finally, building from historical contexts and existing vulnerabilities/risks it is possible for participants to co-create long-term SETS scenarios (40–60 y) to deal with very high uncertainty. The entire process can help inform engineering projects before they are selected and implemented.

Although such planning scenario tools are a valuable addition to inclusive community-based research, there are also a growing number of critical reflections on the capacity of community participants and stakeholders to truly exercise decision-making power within such collaborative processes. If the definition of knowledge ultimately remains in the hands of researchers, technical know-how remains in the hands of engineers, and decisions about projects and financing remain in the hands

20 See, for example, https://data.urexsrn.net/governance-analysis/

of policy-makers, how much power do community actors really have? Building on Sherry Arnstein's 'Ladder of citizen participation' (1969), Hendricks et al. argue for the use of community-based science to make participation more meaningful and effective. Rather than weak participation or tokenism, wherein communities just inform or consult on decisions, a higher rung of the ladder is PAR, in which "local knowledge forms the basis for research and planning" (Cornwall and Jewkes, 1995, p. 1667). PAR involves researchers and stakeholders "in a process through which they are empowered to take charge of the research process and to organize to implement potential solutions or to take action" (Cornwall and Jewkes, 1995, p. 1671).

Hendricks et al. argue that robust PAR, feminist research, and community-based citizen science can foster the co-production of knowledge in ways that empower marginalized groups and recognize differing forms of valuation and risk assessment. The ladder of citizen participation therefore reaches its apex with true community control and ownership of scientific knowledge and decision making:

> Residents, if engaged through community science around infrastructure and hazard issues, have the opportunity to participate in the creation of knowledge and a grounded fact base from which decision-making around policy development, tax allocations, information sharing, program operation, etc. occurs. Community science represents the potential for redistribution of power dynamics in traditional top-down science similar to the redistribution of power that allows communities to have a voice in future government decision-making (Haklay 2013).[21]

The key element in all of these approaches is the integration of social knowledge and dialogical co-learning processes into community understandings of risk, adaptation, and transformation, so that people can take action themselves. The use of scenarios and narrative planning approaches allows for better communication processes between 'experts' and 'local people', or communities, especially if grounded in a local language (rather than the language of expertise). Creating such shared knowledge-networks can bring actors of different types together in a commonly shared framework. This can then support the co-design of engineering interventions informed by that two-way process, with greater likelihood that the involved community participants will understand, maintain, and accept the engineering intervention since they are involved in its co-design. The PAR approach using citizen science to give communities power to co-create both knowledge and action within planning and decision-making processes can furthermore embed local epistemes and forms of valuation into structures of governance and decision-making, supporting not only procedural justice but also epistemic justice.[22]

---

21 Hendricks et al. "Moving Up the Ladder in Rising Waters", p. 3, citing Haklay, "Neogeography and the delusion of democratisation".

22 See Sheller, *Mobility Justice*, on different models of justice within systems of transportation and mobility regimes.

However, it is not always easy for researchers, engineers, or development practitioners to give up control over their own processes and ontological worldview, and to demote their own technical expertise to just one part of co-learning and co-design process. This requires a more modest and critically self-reflexive approach not only to community stakeholder engagement, but also to the kinds of knowledge claims that are made from a position of expertise. This leads me to some reflections on my own experiences in an interdisciplinary research team that was working on participatory engineering in post-earthquake Haiti in 2010-2012, as well as my subsequent return to the area of the Haitian-Dominican border during a second period of research in 2012-2014.[23]

Learning from Post-earthquake Reconstruction in Haiti

Engineering is always embedded in social processes. Community participation can be mobilized to assist in making more sustainable connections not just between technical elements, but also between various stakeholders within any given socio-technical context, including engineers, builders, managers, policy-makers, government officeholders, community-based organizations (CBOs) and citizens. Yet there are still substantial gaps in the achievement of such laudatory goals. In developing a participatory model of engineering in prior work in post-earthquake Haiti, my NSF-supported research team sought to show how more robust social interactions between various stakeholders (including engineers and foreign aid agencies as well as local citizens and government agencies) could improve outcomes in the dynamic environmental, political, and socio-economic constraints and opportunities of a post-disaster transition from emergency response to rebuilding.

Engaging directly in this kind of complex post-disaster coordination of participatory processes in contexts like post-earthquake Haiti is helpful for understanding potential pathways for sustainable development engineering more generally. We asked: how could participatory engineering help to shape a social learning process that influences how effective, equitable and sustainable a particular disaster recovery or transitional development scenario will be? We found that engineers and social scientists working in multi-stakeholder arenas need to pay greater attention to interactions amongst actors, unfolding decision-making processes, and relations of power that in-

23 National Science Foundation (NSF) "Haiti-RAPID: Supporting Haitian Infrastructure Reconstruction Decisions with Local Knowledge" (PI Montalto, Co-PIs M. Piasecki, P. Gurian, M. Sheller), ENG-CBET No. 1032184, 2010-2012; NSF-RAPID: Understanding Sudden Hydro-Climatic Changes and Exploring Sustainable Solutions in the Enriquillo Closed Water Basin SW Hispaniola (PI J. Gonzalez, CCNY, Co-PIs M. Sheller et al.) ENG-CBET No. 1264466, 2012-2013.

form decision-making, governance, and funding allocation processes. The entire 'emergency' social context is striated by uneven relations of power among dozens of international NGO's working in the same location, multiple local, national and international government agencies working at various levels, the actions of local elites including those returning from the diaspora, and the many internal power relations of class, gender, age, etc., within the local social context.

PAR can use deliberative workshops, citizen science tools and co-design and co-management of infrastructure systems to help support more equitable and robust participation, but still faces a number of limitations. Foreign research teams and engineers, as well as humanitarian actors, are deeply enmeshed in structures of power that are avoidable and must be confronted head on. In Haiti, I became aware of the way in which our team benefited from access to 'mobility capital' in the form of rented vehicles, mobile communication technology, and easy access to water, electricity and travel. I also noted that much of the earthquake-response was not being conducted in the local language (in this case Haitian Creole, rather than the general use of French or English by foreigners and educated elites) which excluded many local communities. It was evident that many 'humanitarian' projects were being foisted on local communities, rather than being selected and designed by them, leading to rapid failure and abandonment of such interventions.[24]

Ideally, more democratic and participatory tools can enable 'co-construction of meaning and the sharing of information and understanding regarding a particular context that is to be managed', generating an ongoing deliberative process of 'social learning'.[25] Public deliberations encourage reflexivity about preferences and opinions, and clearer articulation of implicit values, beliefs, interests and concerns, which can inform engineering decisions.[26] A participatory process known in French as 'accompagnement' [accompanying] can help to support experiential learning about the issue at hand, about solutions to technical problems, and about other stakeholders, and can thus support the kind of collective learning or organizational learning that is necessary to effective participatory engineering.[27]

GRET, a French NGO with a long-term Haitian staff working specifically in the water sector in Port-au-Prince, for example, concluded that "Civil society must be in a position to be involved in the reconstruction of the country. In each neighborhood, there should be support for legitimate grassroots organizations and for the water

24 For a full account see Sheller, *Island Futures*.

25 Alexander et al., "Water Needs Assessment"; Etienne et al., "ARDI"; Greenwood and Levin, *Introduction to Action Research*; Pahl-Wostl, "Towards Sustainability in the Water Sector"; Pahl- Wostl, "Participative and Stakeholder Based Policy Design Evaluation"; Pahl-Wostl, "A Conceptual Framework."

26 Parkinson, *Deliberating in the Real World*; Wagenaar, *Meaning in Action*.

27 Daré et al., *Repères méthodologiques pour la mise en oeuvre d'une démarche de modélisation d'accompagnement*; Etienne, "La modélisation d'accompagnement."

committees [komite dlo] so that they can have a say in the matter both around current debates and future actions; and there should be assistance for this local civil society to think and act through [their involvement in] local development projects."[28] Working with CAMEP, the public water authority in Port-au-Prince, GRET had partnered in developing community-run water distributions systems, and gradually expanded them over two decades. Their approach involves making investments back into the community, and building what they called 'social infrastructure' along with physical infrastructure. Social infrastructure is crucial to making physical infrastructure work. Certainly within broader studies of disaster recovery and infrastructure the importance of 'human infrastructure' (Klinenberg, 2015, 2018) and the building of 'infrastructural commons' (Frischmann, 2013) are increasingly recognized as crucial, and should be incorporated within development engineering as well.

How then do you build an infrastructural commons that mobilizes human infrastructure as much as the building of pipes, sewers, and water-treatment systems? In Port-au-Prince, the komite dlo "constituted themselves by integrating the leaders of all primary organizations in the community (i.e., political parties, churches, youth groups, women's associations, networks of notables, etc.). They manage the finances of the water sales and smooth occasional conflicts". In other words, they helped build social networks, social trust, and representation of various constituent groups within the community. Moreover, "With a margin obtained from selling water, they finance small collective infrastructure projects (playgrounds, meeting rooms, footbridges, wastewater channels, bathing facilities, etc.), enhancing and reinforcing their legitimacy. They usually also receive training in financial and accounting techniques."[29] The engineering of the water project, in other words, becomes a vehicle for capacity building for social needs beyond water alone, and creates social trust by investing back into the community.

As we concluded from our research on participatory engineering, "NGO strategies, while helpful during the immediate phase of post-disaster response are often criticized for creating dependencies that cannot be supported in the long term. Even more importantly, however, in their ignorance of local histories of participation and community organization, they may actually be undermining local social organization and 'social capital' that exists amongst CBOs as well as bypassing existing national legal frameworks and policy directives."[30] We argued that participatory engineering could have helped to mediate between the needs of government organizations and foreign NGOs (who want to be seen to be making decisions and taking immediate actions within each UN cluster area), and the needs of community based organizations and earthquake-affected people, who want to have more input into the process. More

28 GRET "CAMEP and Komite dlo dans le cadre du projet Rezodlo," June 2010, p. 11. See also Schuller, "Unstable Foundations," for similar arguments.

29 Ibid.

30 Sheller et al., "Participatory Engineering for Recovery in Post-Earthquake Haiti", p. 21.

equitable participatory processes also create the opportunity to articulate and reflect on the interdependencies between different systems, the epistemic values they incorporate, and the transformative power dynamics they might enact.

In between the temporary measures being implemented by international NGOs in post-earthquake Haiti, and the longer-range plans being developed by government agencies such as DINEPA, there should have been time and space for a middle-range participatory-engineering planning process like that achieved by GRET, which would include building social infrastructure that met multiple needs through participatory community decision-making, co-design, and financing.[31] This means that development engineering needs to create a toolkit of approaches that can be mobilized in various kinds of settings with different temporal framings, ranging from emergency response, to short-term projects, to longer term capacity building and development planning. Each requires different types of community participation, and must pay attention to how that occurs, how it is supported, and how it is built into the process in an ongoing way that is attentive to power differentials and social processes for the construction of knowledge, expertise, and decision-making.

Conclusions

The principles of 'just recovery' suggest that there are better ways to go about reconstruction after natural disasters, whether caused by earthquakes, by hurricanes, or by climate change. Lessons learned about such disaster recovery processes have much to teach us about more equitable development engineering in wider contexts. In many cases practitioners and researchers have begun to recognize that disaster recovery requires greater attention to uneven power relations. The strongest projects build social infrastructure in addition to rebuilding physical infrastructure, using self-governed, decentralized, resilient, community-led infrastructuring processes that can integrate water, energy, food, health, education, and social protection. These participatory action processes are indicative of lessons that might be more broadly applied to development engineering.

Radical planning centers collective agency at the heart of the process. Jacobs argues that "By failing to center community knowledge, avoiding the naming of systemic oppression and not making an explicit commitment to grassroots change, social vulnerability cannot comprehensively articulate and address the issues facing communities that experience the brunt of the effects of disasters."[32] In the practice of 'radical planning', small changes can be key to "creating a sense of collective agency and

31 Ibid.

32 Jacobs, "Black feminism and radical planning", p. 30.

the social and political spaces for radical action".[33] The same goes for ethical engineering: it needs to make a commitment to grassroots change, radical inclusion, and support of collective action, which make projects more sustainable as well as more just. Small steps toward collective inclusion can pave the way for wider transformation in how engineering is conceived and practices. At a minimum this must include use of vernacular language, mobilization of existing community based organizations, financial compensation of participants, inclusive processes, and epistemic bridging before decisions are taken and projects launched. Beyond that, as Jacobs argues, it calls for naming structural inequalities, systemic oppression, committing to societal change and advocating reparative justice for those who may have been harmed by past infrastructural decisions and inequitable built systems.

Researchers and practitioners of development engineering should be trained to support new approaches to what we do on the ground, as well as how we educate engineers. Urgent and complex global challenges demand that we move more quickly to change the way that we engage in participatory engineering, build resilience, and implement co-learning and co-design with community partners. Integrating social sciences and ethical approaches into engineering education are a crucial starting point. Carrying this forward into the field and rethinking how engineering includes and empowers local community partners is the next step to making real progress in areas such as disaster recovery, sustainable development, and climate adaptation. The anticipated outcome will be more engaged, cooperative, and adaptive knowledge-communities that bridge diverse places, encompass pluriversal cultures, and support local forms of knowledge-making to create what will ultimately become more resilient social, ecological, and technological systems.

---

33 Beard, "Learning radical planning," p. 28.

Rob Krueger

Coda: Toward a New Canon of Engineering and Development

"Now back to me."
A meditation for the self-absorbed

"Deep inside I know that trying to fathom things out leads to blindness, that the desire to understand has a built-in brutality that erases what you seek to comprehend. Only experience is sensitive. But perhaps I'm both weak and brutal. I've never been able to resist trying . . ."
Peter Høeg, *Miss Smilla's Feeling for Snow*, 269

"I see you"
Zulu greeting

The quotes above capture the continuum on which this collection of essays rest. For 500 years through the colonial project 'enlightened' humans sought to transform the world so that it represented the hegemon's understanding of how things work. Or, normatively speaking, create a world as it ought to be: either from Judeo-Christian Tradition, Enlightenment thinking, or Western Science and Modernization. The logic roughly goes like this: only through the image/narrative of the master subject could this ambitious feat be realized. Being the other meant 'difference' (not diversity) and, for some, dehumanization. If one fell into this latter category, dehumanization translated into and existence outside the moral community. For subjects existing outside the moral community meant this category was a 'means to an end', a tool of instrumental rationality.

'Now back to me' has been foundational to western ideology for many centuries. Self-assertion and self-promotion arguably since the Greeks introduced solipsism, which was reinforced by Descartes, and promoted in the last century by Fredrich Nietzsche and Ayn Rand. The development project has experienced tensions from this ideology in new ways since the 1950s. Indeed, the Marshall Plan demonstrated that science, technology, markets, and trade could liberate (some of) Europe from the grips of fascism and with it, racism. The United States saw greater opportunity for this kind of scheme; as Truman remarked in his inauguration speech of 1949, the United States could lead an effort to bring a Marshall Plan, or modernization, to the world.

Over the past 75 years, the modernization project sought to bring greater human dignity to those in the majority world, but despite their numbers they remained categorized as a means to an end. During this time, 'aid from the American people' meant your geography had strategic importance needed to win the Cold War. Multilateral donors targeted a geography for modernization if it was strategically placed vis-à-vis Soviet advancement or if it had the raw materials strategists believed were required

https://doi.org/10.1515/9783112218716-015

to win the Cold War – Vietnam and Chile provide examples, respectively. In this role, nation-states of interest were also customers, exchange partners, and borrowers. As one member of the Marx family once offered, "Give a man to fish and he eats for a day. Teach a man to fish and you have ruined a perfectly good business opportunity." Development as modernization distinguishes itself from the colonial project in that experts and expertise (economic, scientific, and technological), possessed by the West, and could be transferred to the global majority, thereby bringing them into the political fold.

The modernization paradigm provided a rewarding platform for instrumental rationality. In contrast to colonialism, however, modernization saw an opportunity for non-Western societies to learn 'rational' thought and could ultimately support their own economies and social growth, albeit at a certain interest rate. Still, we cannot understate the departure of the modernization project from the colonial project. Colonialism represented a project of what one might call 'socialized self-absorption and hubris' where the post-war modernization project could be referred to as 'now back to me'. Modernization was intended to support non-western economies (read: backward), but the benefits would be extracted and exported to the global north. Like colonialism, modernization had its own form of built in, if not less direct, form brutality with as it had a utilitarian orientation.

This argument in no way represents and justification for the previous, and continued, colonization of the land and minds of people born in a place with different social norms and unrecognized scientific and technological innovation. It exposes the myth. It offers a lever to grip and to drive forward 'progress'.

Take William Golding's typology from his essay *Thinking as a Hobby*. Here, colonialism would be understood 'grade-three thinking', which is based on feeling, not thought. He writes of his fear of a grade-three thinking mob, of which, he believed, there are a preponderance of in this world. This mob would not thank you for pointing out the contradictions in their thinking or political goals. A level up from this is grade-two thinking. For Golding, grade two thinking moves beyond feeling to thinking; a grade-two thinker understands there are contradictions to their position. Rather than face them, however, they choose to withdrawal with their eyes and ears open. It is, for example, to offer elegant critique without a solution. This leads us to Miss Smilla.

In this book, we have tried to fathom things out, and we do so with our eyes and ears open. We cannot resist trying to change things, knowing, of course, that the project is never finished. In design one starts where you and your collaborators are. With conviction you may advance your agenda in sustainable ways – ways that are social, ecologically, and economically beneficial. Alas, you move to a new place and the work continues, or even begins anew.

The contributors to this volume have tried to reveal a path forward. We began the volume with an all-to-brief history of development practice, especially the roles of science and engineering in this practice. We then introduced the state-of-the-art thinking about transdisciplinary practice, applied ethics, research methods, and co-design. This, we argue requires humility, empathy forward if you will. The problem and the people experience them are viewed here as partners with special expertise that bring new capacities and opportunities. They are no longer problems that need solving. To do this requires that researchers and practitioners embrace and engage the agency of 'the other' to create new collaborations and peer-to-peer partnerships. Beyond methodological practice, we have offered pedagogical insights and experiences as we educate those who will pick this book up.

We have illustrated the brutality of science, engineering, and development, and, then, sought to set hard limits for this tradition of thinking and doing. Science and engineering for development do not arise from technical or STEM training. Science and engineering for development is not a STEM field. STEM fields have their rightful seat at the table, but they should be one among many. They must be augmented by social scientists, creatives, humanists, and the very people who have the capacity to create change – those in the places we seek to support. But engaging people as transactional stakeholders is still not enough.

'I see you,' or, Sawubona, in Zulu, is a greeting that when offered shows the conveyor 'sees' the receiver's humanness – their dignity, their worth, both alone and to the community. To help do this, we have opened up traditional research methods to be democratic processes where the 'researched' becomes an agent for change. They do so because they co-create solutions to problems that they can take ownership of. Ron Eglash calls this 'generative justice', which, for him, can be summarized that the universal right to participate in the generation of value and directly share in its benefits. To be able to accomplish this, the partners must recognize each other's humanity, their blind spots, their foibles, and their value as contributors to problem-solving.

Returning to Golding, we said he offers three grades of thinking. We have introduced you two: grade three and grade two. What is grade-one thinking? Grade-one thinking asks 'what is truth?' A grade-one thinker does not just ask the question, they seek truth out. They engage the world around them. Recognize their place – and sometimes privilege – in it. Despite the essay's title *Thinking as a Hobby*, truth-seeking to a grade-one thinker is not a hobby at all. Hobbies can be diversions, pure entertainment, or amusement. For the grade-one thinker, thinking as hobby would be unsatisfactory if it went no further. The purpose of grade-one thinking, indeed development engineering, is to interpret the world around us and within us, with an empathic decolonial lens, and based on shared understanding, to facilitate change.

The title of this volume is *Rethinking Development Engineering*. It would be a shame if we only thought about it. Our purpose is to prompt change. While we, the editors of this book, have changed through the process of compiling this volume, which began in mid-2020, we will fail if this is the end. We hope the readers of our effort will enjoy the journey.

References

Abbas, R., Pitt, J. & Michael, K. (2021). Socio-technical design for public interest technology. IEEE Transactions on Technology and Society, 2(2), 55–61.

Abrams, D. B. (2006). Applying transdisciplinary reserach strategies to understanding and eliminating health disparities. Health Education & Behavior, 33(4), 515–531.

Ahmed, I. (2014). Multiculturalism in Sweden: Riots Spark Debates About Identity and Policy. Georgetown Journal of International Affairs, 15(1), 98–104.

Albrecht, G., Freeman, S. et al. (1998). Complexity and human health: the case for a transdisciplinary paradigm. Culture, Medicine and Psychiatry, 22, 55–92.

Alcamo, J., (Ed.). (2008). Environmental Futures: The Practice of Environmental Scenario Analysis. Elsevier.

Alexander, K. S., Moglia, M. & Miller, C. (2010). Water Needs Assessment: Learning to Deal with Scale, Subjectivity and High Stakes. Journal of Hydrology, 388(3–4), 251–257.

Alkire, S. & Foster, J. (2011). Counting and multidimensional poverty measurement. Journal of Public Economics, 95(7–8), 476–487.

Altinbilek, D. (2002). The role of dams in development. Water Science and Technology, 45(8), 169–180.

Alvarez, S. E. (2009). Beyond NGO-ization?: Reflections from Latin America. Development, 52(2), 175–184.

AMP Robotics. (n.d.-a). Retrieved June 13, 2022, from https://www.amprobotics.com/

AMP Robotics. (n.d.-b). Retrieved February 25, 2019, from https://www.amprobotics.com/

Anderson, C. (2008). The end of theory: The data deluge makes the scientific method obsolete. Wired Magazine, 16(7), 16–17.

Andersson, H. (2021). Ethnic Enclaves, Self-Employment, and the Economic Performance of Refugees: Evidence from a Swedish Dispersal Policy. International Migration Review, 55(1). 58–83. https://doi.org/10.1177/0197918320912195.

Arendt, H. (1958). The Human Condition. University of Chicago Press.

Arens, E. & Sheldon, R. (1932). Consumer Engineering. A New Technology for Prosperity. New York/London.

Arndt, H. W. (1981). Economic development: a semantic history. Economic Development and Cultural Change, 29(3), 457–466.

Arnstein, S. R. (1969). A ladder of citizen participation. Journal of the American Institute of Planners, 35(4), 216–224. https://doi.org/10.1080/01944366908977225.

Asaro, P. (2016). Hands up, don't shoot!: HRI and the automation of police use of force. Journal of Human-Robot Interaction, 5(3). 55–69. https://doi.org/10.5898/JHRI.5.3.Asaro.

Atanasoski, N. & Vora, K. (2019). Surrogate Humanity: Race, Robots, and the Politics of Technological Futures. Duke University Press.

Atlas^{TM} | Boston Dynamics. (n.d.). Retrieved June 12, 2022, from https://www.bostondynamics.com/atlas

Bainbridge, W. S. & Roco, M. C. (2015). The Era of Convergence. In: W. Bainbridge & M. Roco (Eds.), Handbook of Science and Technology Convergence. Cham: Springer. https://doi.org/10.1007/978-3-319-04033-2_1-1.

Bainbridge, W. S. (2015). Twenty-First Century Society. In: W. Bainbridge & M. Roco (Eds.), Handbook of Science and Technology Convergence. Cham: Springer. https://doi.org/10.1007/978-3-319-04033-2_45-1.

Baker, J. L. (2000). Evaluating the impact of development projects on poverty: A handbook for practitioners. The World Bank.

Balsiger, P. W. (2004). Supradisciplinary research practices: history, objectives and rationale. Futures, 36, 407–421.

Banerjee, A. V. & Duflo, E. (2009). The Experimental Approach to Development Economics. Annual Review of Economics, 1, 151–178.

Banerjee, A., Jacob, S., Kremer, M., Lanjouw, J. & Lanjouw, P. (2005). Moving to Universal Education: Costs and Tradeoffs. MIT Working Paper.

https://doi.org/10.1515/9783112218716-016

Barnes, T. J. & Sheppard, E. (2010). 'Nothing includes everything': towards engaged pluralism in Anglophone economic geography. Progress in Human Geography, 34(2), 193–214.

Barreteau, O., Bots, P. & Daniell, K. A. (2010). A Framework for Clarifying "Participation" in Participatory Research to Prevent its Rejection for the Wrong Reasons. Ecology and Society, 15(2), 1. http://www.ecologyandsociety.org/vol15/iss2/art1/.

Bashkirova, D., Abdelfattah, M., Zhu, Z., Akl, J., Alladkani, F., Hu, P., Ablavsky, V., Calli, B., Bargal, S. A. & Saenko, K. (2022). ZeroWaste Dataset: Towards Deformable Object Segmentation in Cluttered Scenes. Proceedings of the IEEE/CVF Conference on Computer Vision and Pattern Recognition, 21147–21157.

Basra, R. & Neumann, P. R. (2016). Criminal Pasts, Terrorist Futures: European Jihadists and the New Crime-Terror Nexus. Perspectives on Terrorism, 10(6).

Beard, V. A. (2003). Learning radical planning: The power of collective action. Planning Theory, 2(1), 13–35.

Bebbington, A., Farrington, J., Lewis, D. J. & Wellard, K. (1993). Reluctant Partners? Non-Governmental Organizations, the State and Sustainable Agricultural Development, 1st edn. Routledge.

Benmelech, E. & Klor, E. F. (2018, October). What Explains the Flow of Foreign Fighters to ISIS? Terrorism and Political Violence, 1–24.

Bennett, C. (2012). Supporting the Posts in Development Discourse: Under-development, Over-development, Post-development. Sociology Compass, 6(12), 974–986.

Bennett, T., Savage, M., Silva, E. B., Warde, A., Gayo-Cal, M. & Wright, D. (2009). Culture, Class, Distinction. Routledge.

Berg, J. E., Dickhaut, J. W. & McCabe, K. A. (1995). Trust, reciprocity, and social history. Games and Economic Behavior, 10(1), 122–142.

Bernstein, R. J. (1987). I. One Step Forward, Two Steps Backward: Richard Rorty on Liberal Democracy and Philosophy. *Political Theory*, *15*(4), 538–563.

Bertrand, M., Duflo, E. & Mullainathan, S. (2004). How much should we trust differences-in-differences estimates? Quarterly Journal of Economics, 119(1), 249–275.

Beshah, T. & Hill, S. (2010, March). Mining road traffic accident data to improve safety: role of road-related factors on accident severity in Ethiopia. In 2010 AAAI Spring Symposium Series.

Bewket, W. & Sterk, G. (2002). Farmers' Participation in Soil and Water Conservation Activities in the Chemoga Watershed, Blue Nile Basin, Ethiopia. Land Degradation & Development, 13(3), 189–200.

BigDog – ROBOTS: Your Guide to the World of Robotics. (n.d.). Retrieved June 12, 2022, from https://robots.ieee.org/robots/bigdog/

Bijker, W. E., Hughes, T. P. & Pinch, T. (1987). The Social Construction of Technological Systems. Cambridge: MIT Press.

Billard, A. & Kragic, D. (2019). Trends and challenges in robot manipulation. Science, 364(6446). https://doi.org/10.1126/SCIENCE.AAT8414/ASSET/7AFABCBB-DE33-41EC-BFE5-57A773BEDDB6/ASSETS/GRAPHIC/364_AAT8414_F7.JPEG.

Birk, A. (2021). Seeing through the forest and the trees with drones. Science Robotics, 6(55). 3947. https://doi.org/10.1126/SCIROBOTICS.ABJ3947/ASSET/220EC4FF-E80C-4BE2-9E83-3EBA86C413BD/ASSETS/GRAPHIC/ABJ3947-F1.JPEG.

Blumenstock, J. E. (2016, August). Fighting poverty with data. Science, 353(6301), 753–754.

Bonilla, L. (2021). Perfil Migratorio Panamá 2019-2020. Perfil Migratorio Panamá 2019–2020. https://publications.iom.int/es/node/3636.

Borenstein, J. (2011). Robots and the changing workforce. AI and Society, 26(1), 87–93. https://doi.org/10.1007/s00146-009-0227-0.

Borevi, K. (2013). The Political Dynamics of Multiculturalism in Sweden. In R. Taras (Eds.), Challenging Multiculturalism. European Models of Diversity. Edinburgh University Press, 138–160.

Borevi, K. (2014). Multiculturalism and welfare state integration: Swedish model path dependency. Identities, 21(6), 708–723.

Borjas, G. J. (1995). Ethnicity, neighborhoods, and human-capital externalities. The American Economic Review, 85(3). 365–390. https://www.jstor.org/stable/2118179.
Bosshard, P. (2013). The World Bank is bringing back big, bad dams. The Guardian.
Bourguignon, F. & Chakravarty, S. (2003). The measurement of multidimensional poverty. The Journal of Economic Inequality, 1(1), 25–49.
Bridger, J. C. & Luloff, A. E. (1999). Toward an interactional approach to sustainable community development. The Journal of Rural Studies, 15, 377–387.
Brown, C., McCauley, S. M., & Adler-Abramo, R. (June 2024). Inclusivity Overlooked: A Case Study of People with Disabilities' Inclusion in Climate Resilience Planning in Massachusetts. Journal of Environmental Justice. https://doi.org/10.1089/env.2023.0084.
Brown, C., & Laing, T. (October 2025). Support Networks for Doctoral Students at HBCUs: Insights and Implications. In E. McPherson & T. Laing (Eds.), (In)visible Student Support Networks at HBCUs and PWIs for Black Undergraduate and Graduate Students. SUNY Press, New York. https://sunypress.edu/Books/B/Black-Student-Support-Networks.
Buchanan, R. (1992). Wicked problems in design thinking. Design issues, 8(2), 5–21.
Bustinza, I., Arce, S., Ortiz, A., & McCarthy, M. (2022). Recent Migration Trends in the Americas. International Organization for Migration (IOM) Regional Office for South America. https://rosanjose.iom.int/sites/g/files/tmzbdl1446/files/documents/recent-migration-trends-in-the-americas_en.pdf.
Callaghan, E. G. & Colton, J. (Dec. 2008). Building sustainable & resilient communities: a balancing of community capital. Environment, Development and Sustainability, 10(6), 931–942. doi: 10.1007/s10668-007-9093-4.
Callon, M. (1987). Society in the Making: The Study of Technology as a Tool for Sociological Analysis. In W. E. Bijker, T. P. Hughes & T. Pinch (Eds.), The Social Construction of Technological Systems. Cambridge: MIT Press.
Camerer, C. F. (2003). Behavioral Game Theory. Princeton, NJ: Princeton University Press.
Carbon War Room – RMI. (n.d.). Retrieved June 13, 2022, from https://rmi.org/carbon-war-room/
Carr, E. (2011). Delivering Development: Globalization's Shoreline and the Road to a Sustainable Future. New York: Palgrave.
Carr, E. R. (2012). Small farmers, big impacts. Frontiers in Development, 92–97.
Chamberlin, E. H. (1948). An experimental imperfect market. Journal of Political Economy, 56, 95–108.
Chambers, R. (1983). Putting the Last First: Routledge, 1st edn. Routledge.
Chambers, R. (1986). Putting the Last First. Routledge.
Chambers, R. (1994). The origins and practice of participatory rural appraisal. World Development, 22(7), 953–969.
Chambers, R. (1997). Whose Reality Counts, vol. 25. London: Intermediate Technology Publications.
Chang, H. J. (2003). Rethinking Development Economics. Anthem Press.
Chang, H., Pallathadka, A., Sauer, J., Grimm, N., Zimmerman, R., Cheng, C., Iwaniec, D. M., Kim, Y., Lloyd, R., McPhearson, P. T., Rosenzweig, B., Troxler, T., Welty, C., Brenner, R. & Herreros-Cantis, P. (May, 2021). Assessment of urban flood vulnerability using the social-ecological-technological systems framework in six US cities. Sustainable Cities and Society, 68, 102786.
Chang, H.-J. (2002). Kicking Away the Ladder: Development Strategy in Historical Perspective. London: Anthem.
Chmela-Jones, K. A. (2015). The ethics of Ubuntu and community participation in design. In Proceedings of the 7th International DEFSA Conference. Vanderbijlpark, South Africa, 43–50.
Choudry, A. & Kapoor, D. (2013). NGOization: Complicity, Contradictions and Prospects, 1st edn. Bloomsbury Publishing.
Clark, A. (2003). Natural-Born Cyborgs: Minds, Technologies, and the Future of Human Intelligence. New York: Oxford University Press. Collaborative Resilience.
Collier, S. J. (2017). Neoliberalism and Rule by Experts. In: V. Higgins & W. Larner (Eds.), Assembling Neoliberalism: Expertise, Practices, Subjects. Albury, Australia: Charles Sturt University. 23–44.

Colonia, C. J. G. (2017). Brujería, minería tradicional y capitalismo transnacional en los Andes colombianos: el caso del pueblo minero de Marmato. Instituto Colombiano de Antropología e Historia.
Conrad, J. (2009). The Heart of Darkness. Abingdon: Oxon.
Cooke, B., & Kothari, U. (Eds.). (2001). Participation: The new tyranny?. Zed books.
Coppinger, V. M., Smith, V. L. & Titus, J. A. (1980). Incentives and behavior in English, Dutch, and sealed-bid auctions. Economic Inquiry, 18, 1–22.
Cornwall, A. & Jewkes, R. (1995). What is participatory research? Social Science & Medicine, 41(12), 1667–1676.
Cowls, J., Tsamados, A., Taddeo, M. & Floridi, L. (2021). A definition, benchmark and database of AI for social good initiatives. Nature Machine Intelligence, 3(2), 111–115.
Craggs, R. & Neate, H. (2019). Post-colonial careering and the discipline of geography: British geographers in Nigeria and the UK, 1945–1990. Journal of Historical Geography, 66, 31–42.
Craggs, Ruth, and Hannah Neate. 2019. "Post-colonial careering and the discipline of geography: British geographers in Nigeria and the UK, 1945–1990." *Journal of Historical Geography* 66: 31–42.
Craig, G. & Mayo, M., (Eds.). (1995). Community Empowerment: A Reader in Participation and Development. Zed Books.
Creighton, J. L. (2005). The Public Participation Handbook: Making Better Decisions Through Citizen Involvement. San Francisco: John Wiley & Sons.
Critical Theory. Selected Essays Max Horkheimer, published by Continuum 1982.
Croson, R. & Shang, J. (2008). The impact of downward social information on contribution decisions. Experimental Economics, 11(3), 221–233.
Cross, N. (1982). Designerly ways of knowing. Design Studies, 3(4), 221–227.
Cross, N. (2007). From a design science to a design discipline: Understanding designerly ways of knowing and thinking. In: Design Research Now. Birkhäuser Basel, 41–54.
Cutts, B. B., White, D. D. & Kinzig, A. P. (2011). Participatory geographic information systems for the co-production of science and policy in an emerging boundary organization. Environmental Science and Policy, 14(8), 977–985.
Daly, J. C. (2015). Sweden grapples with rising extremism. Terrorism Monitoring, 13(3).
Daniell, K. A. (2012). Co-Engineering and Participatory Water Management: Organisational Challenges for Water Governance. Cambridge: Cambridge University Press.
De la bellacasa, M. P. (2017). Matters of Care: Speculative ethics in more than human worlds. University of Minnesota Press.
De-Arteaga, M., Herlands, W., Neill, D. B. & Dubrawski, A. (2018). Machine learning for the developing world. ACM Transactions on Management Information Systems (TMIS), 9(2), 1–14.
Della Santina, C., Katzschmann, R. K., Bicchi, A. & Rus, D. (2018). Dynamic control of soft robots interacting with the environment. 2018 IEEE International Conference on Soft Robotics, RoboSoft 2018, 46–53. https://doi.org/10.1109/ROBOSOFT.2018.8404895.
Ding, Y., Xin, B. & Chen, J. (2021). A Review of Recent Advances in Coordination Between Unmanned Aerial and Ground Vehicles. 9(2), 97–117. https://doi.org/10.1142/S2301385021500084.
DiSalvo, C. (2015). Adversarial Design. MIT Press.
Dodson, L. L. & Bargach, J. (2015). Harvesting fresh water from fog in rural Morocco: research and impact Dar Si Hmad's Fogwater Project in Aït Baamrane. Procedia Engineering, 107, 186–193.
Dodson, L. L. (2014). A Foggy Desert: Equitable Information Flow for a Fogwater System in Southwest Morocco. University of Colorado at Boulder.
Does America Need More Innovators?
Dorigo, M., Theraulaz, G. & Trianni, V. (2020). Reflections on the future of swarm robotics. Science Robotics, 5(49). https://doi.org/10.1126/SCIROBOTICS.ABE4385.
Duflo, E., Dupas, P., Kremer, M. & Sinei, S. (2006). Education and HIV/AIDS prevention: Evidence from a randomized evaluation in western Kenya. World Bank Policy Research Working Papers.

Duflo, Esther, Rachel Glennerster, and Michael Kremer (2007). "Using Randomization in Development Economics Research: A Toolkit." *Handbook of Development Economics*, vol. 4, 3895–3962. Elsevier.

Duflo, E., Hanna, R. & Ryan, S. P. (2012). Incentives work: Getting teachers to come to school. American Economic Review, 102(4), 1241–1278.

Dunbabin, M. & Marques, L. (2012). Robots for environmental monitoring: Significant advancements and applications. IEEE Robotics and Automation Magazine, 19(1). 24–39. https://doi.org/10.1109/MRA.2011.2181683.

Dunne, A. & Raby, F. (2013). Speculative Everything: Design, Fiction, and Social Dreaming. MIT Press.

Dunne, A. & Raby, F. (2013). Speculative Everything: Design, Fiction, and Social Dreaming. MIT Press.

EAD. (2021). The IEEE Global Initiative on Ethics of Autonomous and Intelligent Systems. Ethically Aligned Design: A Vision for Prioritizing Human Well-Being with Autonomous and Intelligent Systems, 2nd edn. IEEE. https://standards.ieee.org/wp-.content/uploads/import/documents/other/ead_v2.pdf. Last Accessed on June 12, 2023.

Earthquake Engineering Research Institute (EERI). (2010). The 12 January 2010 Haiti Earthquake: Emerging Research Needs and Opportunities: Report from the Haiti Workshop Held September 30th to October 1st, 2010. Oakland, CA: Earthquake Engineering Research Institute for the National Science Foundation.

Easterly, W. (2006). The White Man's Burden: Why the West's Efforts to Aid the Rest Have Done So Much Ill and So Little Good. New York: Penguin Books.

Easterly, W. (2014). The Tyranny of Experts: Economists, Dictators, and the Forgotten Rights of the Poor. Basic Books.

Edmondson, A. C. (1999). Psychological safety and learning behavior in work teams. Administrative Science Quarterly, 44(2), 350–384.

Eglash, R. & Foster, E. (2017). On the politics of generative justice: African traditions and maker communities. What Do Science, Technology, and Innovation Mean from Africa, 117–135.

Eglash, R. & Foster, E. K. (2017). On the politics of generative justice: African traditions and maker communities. What do sciences, technology, and innovation mean for Africa.

Eglash, R. (1997). African Fractals: Indigenous Design and Modern Computer Engineering. New Brunswick: Rutgers University Press.

Eglash, R. (1997). African Fractiles. University of Minnesota Press.

Eisenberg, E. M. (1984). Ambiguity as strategy in organizational communication. Communication Monographs, 51(3). 227–242. https://doi.org/10.1080/03637758409390197.

Elson, D., (Ed.). (1995). Male Bias in Development. Manchester University Press.

El-Tayeb, F. (2018). Oppressed Majority: Violence and Muslim Communities in Multicultural Europe. In D. Sohail & R. Junaid (Eds.), With Stones in Our Hands: Writings on Muslims, Racism, and Empire 2018. Minneapolis; London: University of Minnesota Press, 83–100.

Elwood, S. (2006). Critical issues in participatory GIS: Deconstructions, reconstructions, and new research directions. Transactions in GIS, 10(5), 693–708.

Engineering, Design, and Society < Colorado School of Mines. https://catalog.mines.edu/undergraduate/programs/edns/#courseinventory (accessed Feb. 01, 2020).

Escobar, A. (1995). Encountering development. Princeton, NJ: Princeton University Press.

Escobar, A. (2011). Encountering Development: The Making and Unmaking of the Third World. Princeton University Press.

Escobar, A. (2018). Designs for the Pluriverse: Radical Interdependence, Autonomy, and the Making of Worlds. Duke University Press.

Escobar, A. (2018). Designs for the Pluriverse: Radical Interdependence, Autonomy, and the Making of Worlds. Duke University Press.

Étienne, M., (Ed.). (2010). La modélisation d'accompagnement: une démarche participative en appui au développement durable [Accompanying Modelling: A Participatory Approach Supporting Sustainable Development]. Versailles, France: Quae éditions.

Étienne, M., Du Toit, D. R. & Pollard, S. (2011). ARDI: A Co-construction Method for Participatory Modeling in Natural Resource Management. Ecology and Society, 16(1), 44.

Federal Deposit Insurance Corporation, D. o. R. a. S. (1997). The LDC Debt Crisis. In: Chap. 5 in History of the Eighties – Lessons for the Future, Volume I: An Examination of the Banking Crises of the 1980s and Early 1990s. Washington, DC: Federal Deposit Insurance Corporation.

Feldman, M. (1991). The meanings of ambiguity: Learning from stories and metaphors. In: P. Frost, L. Moore, M. Reis Louis, C. Lundberg & J. Martin (Eds.), Reframing Organizational Culture. Newbury Park, CA: Sage Publications.

Ferry, E. E. & Ferry, S., La Batea: Impressions of Gold in Colombia. Red Hook Editions, 2017.

Fisher, R. A. (1926). The arrangement of field trials. Journal of the Ministry of Agriculture of Great Britain, 33, 503–513.

Fjeld, J., Achten, N., Hilligoss, H., Nagy, A. & Srikumar, M. (2020). Principled Artificial Intelligence: Mapping Consensus in Ethical and Rights-Based Approaches to Principles for AI. Berkman. Klein Center Research Publication.

Forster, T., Kentikelenis, A. E., Reinsberg, B., Stubbs, T. H. & King, L. P. (2019). How structural adjustment programs affect inequality: A disaggregated analysis of IMF conditionality, 1980–2014. Social Science Research, 80, 83–113.

Fortmann, L. (1985). Seasonal dimensions of rural social organisation. The Journal of Development Studies, 21(3), 377–389.

Foster, J., Greer, J. & Thorbecke, E. (1984). A class of decomposable poverty measures. Econometrica: Journal of the Econometric Society, 761–766.

Fountain, H. "Managing Disasters with Small Steps." The New York Times, 2010. http://www.nytimes.com/2010/01/19/science/19reli.html?pagewanted=all&_r=0 (accessed September 05, 2014).

French, H. W. (2014). Spare the Advice. New York Times book Review: New York Times.

Frischmann, B. M. (2013). Infrastructure: The Social Value of Shared Resources. Oxford and New York: Oxford University Press.

Fry, T. (1999). A New Design Philosophy: An Introduction to Defuturing. Sydney, Australia: UNSW Press.

Fry, T. (2004). The voice of sustainment: design ethics as futuring. Design Philosophy Papers, 2(2), 145–156.

Fry, T. (2007). Redirective practice: an elaboration. Design Philosophy Papers, 5(1), 5–20.

Fry, T. (2011). Urban futures in the age of unsettlement. Futures, 43(4), 432–439.

Fry, T. (2017). Design for/by "the Global South". Design Philosophy Papers, 15(1), 3–37.

Funtowicz, S. O. & Ravetz, J. R. (1993). Science for the post-normal age. Futures, 25(7), 739–755.

Furlong, K. (2013). The dialectics of equity: consumer citizenship and the extension of water supply in Medellin, Colombia. Annals of the Association of American Geographers, 103(5), 1176–1192.

Galada, H. C., Gurian, P. L., Montalto, F., Sheller, M., Piasecki, M., Ayalew, T. B. & O'Connor, S. (2014). Attitudes Toward Post-earthquake Water and Sanitation Management and Payment Options in Léogâne, Haiti. Water International, 38(6), 744–757.

Galada, H. C., Montalto, F. A., Gurian, P. L., Sheller, M., Ayalew, T. B. & O'Connor, S. (2014). Assessing Preferences Regarding Centralized and Decentralized Water Infrastructure in Post-earthquake Léogâne, Haiti. Earth Perspectives: Transdisciplinarity Enabled, 1(5). doi: 10.1186/2194-6434-1-5.

Galloway, T. A., Gustafsson, B. A., Pedersen, P. & Österberg, T. (2015). Immigrant Child Poverty in Scandinavia: A Panel Data Study. In T. I. Garner & K. S. Short (Eds.), Measurement of Poverty, Deprivation, and Economic Mobility. Bingley, UK: Emerald Group Publishing Limited, 185–219.

Gavin, M. & Rodrik, D. (1995). The World Bank in historical perspective. The American Economic Review, 85(2), 329–334.

Gibbons, M., Limoges, C. et al. (1994). The New Production of Knowledge: The Dynamics of Science and Research in Contemporary Societies. London: SAGE Publications.

Gibson, J., Hendricks, M. D. & Wells, J. C. (2018). From engagement to empowerment: How heritage professionals can incorporate participatory methods in disaster recovery to better serve socially

vulnerable groups. International Journal of Heritage Studies, 25(6), 1–15. https://doi.org/10.1080/13527258.2018.1530291.
Giedeon, S. (1948). Mechanisation Takes Command. Minneapolis: University of Minnesota Press.
Gigone, D. & Hastie, R. (1993). The common knowledge effect: Information sharing and group judgment. Journal of Personality and Social Psychology, 65(5). 959–974. https://doi.org/10.1037/0022-3514.65.5.959.
Glewwe, P., Kremer, M. & Moulin, S. (2009). Many children left behind? Textbooks and test scores in Kenya. American Economic Journal: Applied Economics, 1(1), 112–135.
Goldstein, B. E. (2012). Collaborative Resilience: Moving Through Crisis to Opportunity. MIT Press.
Goodfellow, I. J., Shlens, J., & Szegedy, C. (2014, December). Explaining and harnessing adversarial examples. In 3rd International Conference on Learning Representations (ICLR) 2015.
Goralski, M. A. & Tan, T. K. (2020). Artificial intelligence and sustainable development. The International Journal of Management Education, 18(1), 100330.
Goralski, M. A. & Tan, T. K. (2022). Artificial intelligence and poverty alleviation: Emerging innovations and their implications for management education and sustainable development. The International Journal of Management Education, 20(3), 100662.
Goralski, M. A. & Tan, T. K. (2023). Artificial Intelligence: Poverty Alleviation, Healthcare, Education, and Reduced Inequalities in a Post-COVID World. In The Ethics of Artificial Intelligence for the Sustainable Development Goals. Cham: Springer International Publishing. 97–113.
Greenwood, D. & Levin, M. (1998). Introduction to Action Research: Social Research for Social Change. Thousand Oaks, CA: Sage.
GRET, CAMEP and Komite dlo dans le cadre du projet Rezodlo. Rapport de diagnostic: Situation dans les quartiers défavorisés de Port au Prince 4 mois apres le séisme de 12 janvier 2010 [Diagnostic report: Situation in the disadvantaged districts of Port au Prince four months after the earthquake of 12 January 2010]. Port au Prince: GRET, June 2010.
Grillo, R. D. & Stirrat, R. L. (1997, 2020). Discourses of Development: Anthropological Perspectives, 1st edn. Routledge.
Grimes, M. G. (2018). The pivot: how founders respond to feedback through idea and identity work. Academy of Management Journal, 61(5). 1692–1717. https://journals.aom.org/doi/abs/10.5465/amj.2015.0823.
Guenat, S., Purnell, P., Davies, Z. G., Nawrath, M., Stringer, L. C., Babu, G. R., Balasubramanian, M., Ballantyne, E. E., Bylappa, B. K., Chen, B. & De Jager, P. (2022). Meeting sustainable development goals via robotics and autonomous systems. Nature Communications, 13(1), 3559.
Gueth, Werner, Rolf Schmittberger, and Bernd Schwarze (1982). "An Experimental Analysis of Ultimatum Bargaining." *Journal of Economic Behavior and Organization*, 3(4), 367–388.
Guth, W., Schmittberger, R. & Schwarze, B. (1982). An experimental analysis of ultimatum bargaining. Journal of Economic Behavior and Organization, 3(4), 367–388.
Guzak, K. L., Paterson, K. & Archer, A., "A Comparative Assessment of Graduate versus Undergraduate Student Outcomes via International Community Engagement Programs," Presented at the 2012 ASEE Annual Conference & Exposition, Jun. 2012, pp. 25.27.1–25.27.27, Accessed: Feb. 01, 2020. [Online]. Available: https://peer.asee.org/a-comparative-assessment-of-graduate-versus-undergraduate-student-outcomes-via-international-community-engagement-programs.
Hadorn, G. H., Hoffmann, H., et al, (Eds.). (2008). Handbook of Transdisciplinary Research. Springer.
Hagelund, A. (2020). After the Refugee Crisis: Public Discourse and Policy Change in Denmark, Norway and Sweden. Comparative Migration Studies, 8(1). 13. https://doi.org/10.1186/s40878-019-0169-8.
Haidegger, T., Mai, V., Mörch, C. M., Boesl, D. O., Jacobs, A., Khamis, A., Lach, L. & Vanderborght, B. (2023). Robotics: enabler and inhibitor of the sustainable development goals. Sustainable Production and Consumption, 43, 422–434.
Haklay, M. (2013). Neogeography and the delusion of democratisation. Environmentand Planning A, 45(1), 55–69.
Hall, O., Ohlsson, M., & Rögnvaldsson, T. (2022, October). A review of explainable AI in the satellite data, deep machine learning, and human poverty domain. Patterns, 3(10).

Hall, R., Stevens, R. & Torralba, T. (2005). Disrupting representational infrastructure in conversations across disciplines. In S. J. Derry, C. D. Schunn & M. A. Gernsbacher (Eds.), Interdisciplinary Collaboration: An Emerging Cognitive Science. Mahwah. New Jersey: Lawrence Erlbaum Associates, Publishers.

Hansen, M. F., Schultz-Nielsen, M. L. & Tranæs, T. (2017). The Fiscal Impact of Immigration to Welfare States of the Scandinavian Type. Journal of Population Economics, 30(3). 925–952. https://doi.org/10.1007/s00148-017-0636-1.

Hargadon, A. & Sutton, R. I. (1997). Technology brokering and innovation in a product development firm. Administrative Science Quarterly, 42(4), 716–749.

Harrison, G. W. & List, J. (2004). Field experiments. Journal of Economic Literature, 42, 1009–1055.

Harvey, D. (1989). Condition of Postmodernity. London: Basil Blackwell.

Harvey, S. (2014). Creative synthesis: Exploring the process of extraordinary group creativity. Academy of Management Review, 39(3), 324–343.

Hausman, J. A. & Wise, D. A. (1985). Social Experimentation. Chicago, IL: University of Chicago Press, 1–55.

He, B., Wang, S. & Liu, Y. (2019). Underactuated robotics: A review. 16(4). https://doi.org/10.1177/1729881419862164.

Heffernan, N. T. & Heffernan, C. L. (2014). The ASSISTments ecosystem: Building a platform that brings scientists and teachers together for minimally invasive research on human learning and teaching. International Journal of Artificial Intelligence in Education, 24(4), 470–497.

Hendricks, M., Meyer, M. & Wilson, S. (2022). Moving up the ladder in rising waters: community science in infrastructure and hazard mitigation planning as a pathway to community control and flood disaster resilience. Citizen Science: Theory and Practice, Online first. doi: 10.5334/cstp.462.

Henrich, J., Bowles, R., Boyd, S., Camerer, C., Fehr, E., Gintis, H. & McElreath, R. (2001). In search of homo-economicus: Behavioral experiments in 15 small-scale societies. American Economic Review, 91, 73–78.

Hilbert, M. (2016). Big data for development: A review of promises and challenges. Development Policy Review, 34(1), 135–174.

Hildreth, P. M. & Kimble, C. (2002). The duality of knowledge. Information Research, 8.

Hilson, G. (2002). Small-scale mining and its socio-economic impact in developing countries. Natural Resources Forum, 26(1), 3–13. doi: 10.1111/1477-8947.00002.

Hirschman, A. O. (1964). Development Projects Observed. Brookings Institution Press.

Hirschman, A. O. (1964). Development Projects Observed. New York, NY: Brookings Institution Press.

Holt, C. A. & Laury, S. K. (2002). Risk aversion and incentive effects. American Economic Review, 92(5), 1644–1655.

IEEE (n.d.). IEEE at a glance https://www.ieee.org/about/at-a-glance.html

IEEE Spectrum. (n.d.). Stanford's Humanoid Diving Robot Takes on Undersea Archaeology and Coral Reefs. Retrieved June 13, 2022, from https://spectrum.ieee.org/stanford-ocean-one-humanoid-diving-robot

IEEE Standards. (2021). The IEEE Global Initiative on Ethics of Autonomous and Intelligent Systems. Ethically Aligned Design: A Vision for Prioritizing Human Well-being with Autonomous and Intelligent Systems, Second. IEEE. https://standards.ieee.org/wp-.content/uploads/import/documents/other/ead_v2.pdf.

IEEE Standards. (n.d.) The IEEE Global Initiative on Ethics of Autonomous and Intelligent Systems https://standards.ieee.org/industry-connections/ec/autonomous-systems/

Illich, I. (2003). Medical nemesis. Journal of Epidemiology & Community Health, 57(12), 919–922.

Institute of Scrap Recycling Industries (ISRI). (2018). Recycling Industry Yearbook. 2018.

IOM International Organization for Migration. (2011). Glossary on Migration. In: International Migration Law 25. International Organization for Migration. http://www.epim.info/wp-content/uploads/2011/01/iom.pdf.

Irani, L., Vertesi, J., Dourish, P., Philip, K. & Grinter, R. E. (2010, April). Postcolonial Computing: A Lens on Design and Development. In: Proceedings of the SIGCHI conference on human factors in computing systems. 1311–1320.

Irwin, T., Kossoff, G. & Tonkinwise, C. (2015). Transition design provocation. Design Philosophy Papers, 13(1), 3–11.

Isaac, R. M., Walker, J. M. & Thomas, S. H. (1984). Divergent evidence on free riding: An experimental examination of possible explanations. Public Choice, 43, 113–149.

Iwaniec, D. M., Grimm, N. B., McPhearson, P. T., Berbés-Blázquez, M., Cook, E. M. & Munoz-Erickson, T. A. (2021). A framework for resilient urban futures. In: Z. Hamstead, D. M. Iwaniec, P. T. McPhearson, M. Berbés-Blázquez, E. M. Cook & T. A. Munoz-Erickson (Eds.), Resilient Urban Futures. Springer, 1–9. ISBN: 978-3-030-63130-7.

Jacobs, F. (2019). Black feminism and radical planning: New directions for disaster planning research. Planning Theory, 18(1), 24–39.

Jaeger, B. & LaRochelle, E., "Ewb^2 Engineers Without Borders: Educationally, A World Of Benefits," presented at the 2009 Annual Conference & Exposition, Jun. 2009, pp. 14.597.1–14.597.23, Accessed: Mar. 14, 2020. [Online]. Available: https://peer.asee.org/ewb-2-engineers-without-borders-educationally-a-world-of-benefits.

Jameton, A. (1984). Nursing practice: The Ethical Issues. Englewood Cliffs: Prentice-Hall.

Jantsch, E. (1970). From forecasting and planning to policy sciences. Policy Sciences, 1(1), 31–47.

Jantsch, E. (1972). Inter-and transdisciplinary university: a systems approach to education and innovation. Higher Education, 1(1), 7–37.

Jean, N., Burke, M., Xie, M., Davis, W. M., Lobell, D. B., & Ermon, S. (2016, August). Combining satellite imagery and machine learning to predict poverty. Science, 353(6301), 790–794.

Jiang, N., Liu, X., Wang, S., Yu, X., Yin, S., Duan, S., Wang, S., Zhang, R. & Li, S. (2019). Pollution Characterization, Source Identification, and Health Risks of Atmospheric-Particle-Bound Heavy Metals in PM10 and PM2.5 at Multiple Sites in an Emerging Megacity in the Central Region of China. Aerosol and Air Quality Research, 19(2). 247–271. https://doi.org/10.4209/AAQR.2018.07.0275.

Kala, P. (2001). In the spaces of erasure: globalisation, resistance and Narmada river. Economic and Political Weekly, 1991–2002.

Kapuire, G. K., Cabrero, D. G., Stanley, C. & Winschiers-Theophilus, H. (2015, December). Framing technology design in Ubuntu: two locales in pastoral Namibia. In Proceedings of the Annual Meeting of the Australian Special Interest Group for Computer Human Interaction. 212–216.

Katz, J. (2013). The Big Truck That Went By: How the World Came to Save Haiti and Left Behind a Disaster. New York, NY: Palgrave Macmillan.

Kelman, I., Mercer, J. & West, J. J. (2009). Combining different knowledges: community-based climate change adaptation in small island developing states. Participatory Learning and Action: Community-Based Adaptation to Climate-Change, 60, 41–53.

Kennedy, J., Ashmore, J., Babister, E. & Kelman, I. (2008). The Meaning of 'Build Back Better': Evidence from Post-tsunami Aceh and Sri Lanka. Journal of Contingencies and Crisis Management, 16(1), 24–36.

Kessel, F. & Rosenfield, P. L. (2008). Toward transdisciplinary research: historical and contemporary perspectives. American Journal of Preventive Medicine, 35(2S), S225–S234.

Kettle, N. P., Dow, K., Tuler, S., Webler, T., Whitehead, J. & Miller, K. M. (2014). Integrating scientific and local knowledge to inform risk-based management approaches for climate adaptation. Climate Risk Management, 4–5, 17–31.

Khagram, S. (2018). Dams and development. In: Dams and Development. Cornell University Press.

Khoramshahi, M. & Billard, A. (2019). A dynamical system approach to task-adaptation in physical human-robot interaction. Autonomous Robots, 43, 927–946. https://doi.org/10.1007/s10514-018-9764-z.

Khushf, G. (2004). Systems theory and the ethics of human enhancement: a framework for NBIC convergence. Annals of the New York Academy of Sciences, 1013(1), 124–149.

Kim, Y., Mannetti, L. M., Iwaniec, D. M., Grimm, N. B., Berbés-Blázquez, M. & Markolf, S. (2021). Social, Ecological, and Technological Strategies for Climate Adaptation. In: Z. A. Hamstead, D. M. Iwaniec, T. McPhearson, M. Berbés-Blázquez, E. M. Cook & T. A. Muñoz-Erickson (Eds.), Resilient Urban Futures. The Urban Book Series. Cham: Springer, https://doi.org/10.1007/978-3-030-63131-4_3.

Kitchin, R. (2021). Data Lives: How Data Are Made and Shape Our World. Bristol University Press.

Kitchin, R. (2021). The Data Revolution: A Critical Analysis of Big Data, Open Data and Data Infrastructures.

Klein, J. T., Grossenbacher-Mansuy, W., et al., (Eds.). (2001). Transdisciplinarity: Joint Problem Solving among Science Technology, and Society: An Effective Way for Managing Complexity. Basel: Birkhauser Verlag.

Klinenberg, E. (2015). Heat Wave: A Social Autopsy of Disaster in Chicago. Chicago: University of Chicago Press.

Klinenberg, E. (2018). Palaces for the People: How Social Infrastructure Can Help Fight Inequality, Polarization, and the Decline of Civic Life. New York: Broadway Books.

Koselleck, R. (2000). Critique and Crisis: Enlightenment and the Pathogenesis of Modern Society. MIT Press.

Kremer, Michael, Edward Miguel, and Rebecca Thornton (2004). "Incentives to Learn." Harvard University Working Paper.

Kremer, M., Miguel, E. & Thornton, R. (2009). Incentives to learn. Review of Economics and Statistics, 91(3), 437–456.

Lal, D. (2000). Does modernization require westernization? The Independent Review, 5(1), 5–24.

Laplonge, D. (Apr. 2017). The 'un-womanly' attitudes of women in mining towards the environment. The Extractive Industries and Society, 4(2), 304–309. doi: 10.1016/j.exis.2017.01.011.

Laschi, C., Mazzolai, B. & Cianchetti, M. (2016). Soft robotics: Technologies and systems pushing the boundaries of robot abilities. Science Robotics, 1(1). https://doi.org/10.1126/SCIROBOTICS.AAH3690/SUPPL_FILE/EAAH3690_SM.PDF.

Latour, B. & Woolgar, S. (2013). Laboratory Life: The Construction of Scientific Facts. Sage.

Latour, B. (1991). We Have Never Been Modern. Harvard University Press.

Latour, B. (2005). Reassembling the Social: An Introduction to Actor-Network Theory. Oxford: Oxford University Press.

Latour, B. (2005). Reassembling the Social: An Introduction to Actor-Network Theory. Oxford: Oxford University Press.

Latour, B. (2008). A cautious Prometheus? A few steps toward a philosophy of design (with special attention to Peter Sloterdijk). In: Proceedings of the 2008 annual international conference of the design history society.

Latour, B. (2020). What protective measures can you think of so we don't go back to the pre-crisis production model? http://www.bruno-latour.fr/sites/default/files/downloads/P-202-AOC-ENGLISH_1.pdf

Laufer, W. S. (2003). Social accountability and corporate greenwashing. Journal of Business Ethics, 43, 253–261.

Lawrence, M., Janzwood, S. & Homer-Dixon, T. (2022). What is a global polycrisis. Cascade Institute, Technical Paper, 4.

Fry, T. (1999). A New Design Philosophy: An Introduction to Defuturing. UNSW Press.

Lawson, V. (2014). Making Development Geography, 1st edn. Routledge.

Lawther, P. M. (2009). Community involvement in post disaster re-construction – case study of the British Red Cross Maldives Recovery Program. International Journal of Strategic Property Management, 13(2), 153–169.

Leadership, Organizing and Action – Women & Subsistence Mining in Medellín, Colombia | MIT D-Lab. Accessed: Feb. 01, 2020. [Online]. Available: https://d-lab.mit.edu/news-blog/blog/leadership-organizing-and-action-women-subsistence-mining-medellin-colombia.

LeCun, Y., Bengio, Y. & Hinton, G. (2015). Deep learning. Nature, 521(7553), 436–444.

Lee, J. & Zhou, M. (2014). The Success Frame and Achievement Paradox: The Costs and Consequences for Asian Americans. Race and Social Problems, 6(1). 38–55. https://doi.org/10.1007/s12552-014-9112-7.

Lee, J. & Zhou, M. (2015). The Asian American Achievement Paradox. The Asian American Achievement Paradox. New York, NY, US: Russell Sage Foundation.

Leeuw, F. L. & Vaessen, J. (2009). Impact evaluations and development: NONIE guidance on impact evaluation.

Leon, E., Kelman, I., Kennedy, J. & Ashmore, J. (2009). Capacity building lessons from a decade of transitional settlement and shelter. International Journal of Strategic Property Management, 13(3), 247–265.

Leung, K. & Wang, J. (2015). Social processes and team creativity in multicultural teams: A socio-technical framework. Journal of organizational behavior, 36(7). 1008–1025. https://doi.org/10.1002/job.2021.

Levina, M. & Hasinoff, A. A. (2017). The Silicon Valley ethos: Tech industry products, discourses, and practices. Television & New Media, 18(6), 489–495.

Levitt, S. D. & List, J. A. (2009). Field experiments in economics: The past, the present, and the future. European Economic Review, 53(1), 1–18.

Lewis, W. A. (1954). Economic development with unlimited supplies of labour.

Leydens, J. A. & Lucena, J. C. (2014). Social justice: a missing, unelaborated dimension in humanitarian engineering and learning through service. International Journal for Service Learning in Engineering, Humanitarian Engineering and Social Entrepreneurship, 9(2), 1–28.

Leydens, J. A. & Lucena, J. C. (2017). Engineering Justice: Transforming Engineering Education and Practice. John Wiley & Sons.

Leydens, J. A. & Lucena, J. C. (2017). Engineering Justice: Transforming Engineering Education and Practice. John Wiley & Sons.

Li, T. M. (2007). The Will to Improve Governmentality, Development, and the Practice of Politics. Duke University Press.

libreriadelaU.com, “De oro están hechos mis días.” https://www.libreriadelau.com/de-oro-estan-hechos-mis-dias-hombre-nuevo-editores-9789588245843-novela-y-narrativa/p (accessed Feb. 01, 2020).

Lingo, E. L. & O'Mahony, S. (2010). Nexus work: brokerage on creative projects. Administrative Science Quarterly, 55(1). 47–81. https://doi.org/10.2189/asqu.2010.55.1.47.

Lingo, E. L. (2020). Entrepreneurial leadership as creative brokering: the process and practice of co-creating and advancing opportunity. Journal of Management Studies, 57(5). 962–1001. https://doi.org/10.1111/joms.12573.

Lipton, M. (1977). Why Poor People Stay Poor: A Study of Urban Bias in World Development. London; Canberra, ACT; Temple Smith: Australian National University Press.

Lipton, M. (1984). Urban bias revisited. Journal of Development Studies, 20(3), 139–166.

Lipton, Z. C. (2018). The mythos of model interpretability: In machine learning, the concept of interpretability is both important and slippery. Queue, 16(3), 31–57.

List, J. A. & Shogren, J. (1998). Experimental calibration of the difference between actual and hypothetical reported valuations. Journal of Economic Behavior and Organization, 37(2), 193–205.

Loy, D. (1985). Wei-wu-wei: Nondual action. Philosophy East and West, 35(1), 73–86.

Lucena, J., Schneider, J. & Leydens, J. A. (2010). Engineering and Sustainable Community Development. San Rafael, CA: Morgan & Claypool.

Luebkeman, C. (2015). 2050: Designing Our Tomorrow. John Wiley & Sons.

Lundin, R. A., Arvidsson, N., Brady, T., Ekstedt, E., Midler, C. & Sydow, J. (2015). Managing and Working in Project Society Institutional Challenges of Temporary Organizations. Cambridge: Cambridge University Press.

Lynam, T., De Jong, W., Sheil, D., Kusumanto, T. & Evans, K. (2007). A review of tools for incorporating community knowledge, preferences, and values into decision making in natural resources management. Ecology and Society, 12(1). 5 Article 5.

Macellari, M., Yuriev, A., Testa, F. & Boiral, O. (2021). Exploring bluewashing practices of alleged sustainability leaders through a counter-accounting analysis. Environmental Impact Assessment Review. 86(106489).

Macon, B. et al. (2019). The Future of Work in America: People and Places, Today and Tomorrow. McKinsey Global Institute.

Mai, V., Vanderborght, B., Haidegger, T., Khamis, A., Bhargava, N., Boesl, D. B., Gabriels, K., Jacobs, A., Moon, A., Murphy, R. & Nakauchi, Y. (2022). The role of robotics in achieving the united nations sustainable development goals – the experts' meeting at the 2021 IEEE/RSJ IROS workshop [industry activities]. IEEE Robotics & Automation Magazine, 29(1), 92–107.

Mainemelis, C., Kark, R. & Epitropaki, O. (2015). Creative leadership: a multi-context conceptualization. The Academy of Management Annals, 9(1). 393–482. https://doi.org/10.1080/19416520.2015.1024502.

Malpass, M. (2018). Critical Design in Context: History, Theory, and Practice. Bloomsbury Publishing.

Mamidipudi, A. (2019). Crafting Innovation, Weaving Sustainability: Theorizing Indian Handloom Weaving as Sociotechnology. Comparative Studies of South Asia. Africa and the Middle East, 39(2), 241–248.

Mansuri, G. & Rao, V. (2013). Development: Does Participation Work? World Bank Publication.

Manzini, E. (2015). Design, When Everybody Designs: An Introduction to Design for Social Innovation. MIT Press.

Margolin, V. (2002). The Politics of the Artificial: Essays on Design and Design Studies. University of Chicago Press.

Markolf, S. A., Chester, M. V., Eisenberg, D. A., Iwaniec, D. M., Davidson, C. I., Zimmerman, R., Miller, T. R., Ruddell, B. L. & Chang, H. (2018). Interdependent infrastructure as linked social, ecological, and technological systems (SETSs) to address lock-in and enhance resilience. Earth's Future, 6(12), 1638–1659.

Masolo, D. A., Chirikure, S., Augusto, G., Pype, K., Eglash, R., Foster, E. . . . Daniels, C. (2017). In What Do Science, Technology, and Innovation Mean from Africa? C. Mavhunga (Ed.), MIT Press.

Mattson, C. A. & Wood, A. E. (2014). Nine Principles for Design for the Developing World as Derived from the Engineering Literature. Journal of Mechanical Design, 136(12), 121403/1-121403/15.

Mavhunga, C. (2017). What Do Science, Technology, and Innovation Mean from Africa? The MIT Press.

Mavhunga, C. (2017). What Do Science, Technology, and Innovation Mean from Africa? The MIT Press.

Mavhunga, C. (2018). The Mobile Workshop: The Tsetse Fly and African Knowledge Production. MIT Press.

Mavhunga, C. C. (2014). Transient Workspaces: Technologies of Everyday Innovation in Zimbabwe. Boston, USA: MIT Press.

Mazzurco, A. & Jesiek, B. K. (2017). Five Guiding Principles to Enhance Community Participation in Humanitarian Engineering Projects. Journal of Humanitarian Engineering. 5(2).

McDonell, G. J. (2000). Disciplines as Cultures: Towards Reflection and Understanding. In: M. A. Somerville & D. Rapport, Transdiscplinarity: Recreating Integrated Knowledge. Oxford, UK: EOLSS Publishers Co. Ltd., 25–37.

McDonough, W. & Braungart, M. (2010). Cradle to Cradle: Remaking the Way We Make Things. Farrar, Straus and Giroux.

McEwan, C. (2009). International Encyclopedia of Human Geography. R. Kitchin & N. Thrift (Eds.), Durham University.

McEwan, C. (2018). Postcolonialism, Decoloniality and Development. Routledge.

McFarlane, C., Desai, R. & Graham, S. (2014). Informal urban sanitation: everyday life, poverty, and comparison. Annals of the Association of American Geographers, 104(5), 989–1011.

McGuinness, T. D. & Schank, H. (2021). Power to the Public: The Promise of Public Interest Technology. Princeton University Press.

Mercer, J., Kelman, I., Lloyd, K. & Suchet-Pearson, S. (2008). Reflections on use of participatory research for disaster risk reduction. Area, 40(2), 172–183.

Mercer, J., Kelman, I., Suchet-Pearson, S. & Lloyd, K. (2009). Integrating indigenous and scientific knowledge bases for disaster risk reduction in Papua New Guinea. Geografiska Annaler: Series B, Human Geography, 91(2), 157–183.

Merino, L., Caballero, F., Ramiro Martínez-De-Dios, J., Maza, I., Ollero, A., Caballero, F., Martínez-De-Dios, J. R., Maza, I., Ollero, A., Martínez-De-Dios, J. R. & Ollero, A. (2012). An Unmanned Aircraft System for Automatic Forest Fire Monitoring and Measurement. Journal of Intelligent and Robotic Systems, 65, 533–548. https://doi.org/10.1007/s10846-011-9560-x.

Meyer, M. A., Hendricks, M., Newman, G. D., Masterson, J. H., Cooper, J. T., Sansom, G., Gharaibeh, N., Horney, J., Berke, P., Van Zandt, S. & Cousins, T. (2018). Participatory action research: tools for disaster resilience education. International Journal of Disaster Resilience in the Built Environment, 9(4/5). 402–419. https://doi.org/10.1108/IJDRBE-02-2017-0015.

Mhlanga, D. (2020). Artificial Intelligence (AI) and poverty reduction in the Fourth Industrial Revolution (4IR).

Mhlanga, D. (2021). Artificial intelligence in the industry 4.0, and its impact on poverty, innovation, infrastructure development, and the sustainable development goals: Lessons from emerging economies? Sustainability, 13(11), 5788.

Midheme, E. & Moulaert, F. (2013). Pushing back the frontiers of property: Community land trusts and low-income housing in urban Kenya. Land Use Policy, 35, 73–84.

Miguel, E. & Kremer, M. (2004). Worms: identifying impacts on education and health in the presence of treatment externalities. Econometrica, 72(1), 159–217.

Miller, D. (2001). Home Possessions: Material Culture Behind Closed Doors. Oxford, UK: Berg.

Mitchell, T. (2002). Rule of Experts: Egypt, Techno-politics, Modernity. University of California Press.

Mohai, P., Pellow, D. & Roberts, J. T. (2009). Environmental justice. Annual Review of Environment and Resources, 34, 405–430.

Monteiro, M. (2019). Ruined by Design: How Designers Destroyed the World, and What We Can Do to Fix It. Mule Design.

Morozov, E. (2013). To Save Everything, Click Here: The Folly of Technological Solutionism. Public Affairs.

Moseson et al. (2012) Moseson, A. J., Lama, L. & Tangorra, J. (2015). Development by Technology Seeding. Journal of International Development, 27(4), 489–503.

Moyo, D. & Ferguson, N. (2010). Dead Aid: Why Aid Is Not Working and How There Is a Better Way for Africa. Macmillan.

Muller, M. J., Wildman, D. M. & White, E. A. (1993). Taxonomy of PD practices: A brief practitioner's guide. Communications of the ACM, 36(6), 26–28.

Munger, J. & Van Dael, R. (2020). Putting People at the Heart of Policy Design Using Human-Centered Design to Serve All. Manila, Philippines: Asian Development Bank. https://www.adb.org/publications/people-policy-design-human-centered-design.

Murcott, S. (2007). Co-evolutionary Design for Development: Influences Shaping Engineering Design and Implementation in Nepal and the Global Village. Journal of International Development, 19(1), 123–144.

Naper, A. A. (2022). Compassionate Border Securitisation? Border Control in the Scandinavian News Media during the 'Refugee Crisis.'. Nordic Journal of Migration Research, 12(1). 4–20. https://doi.org/10.33134/njmr.494.

Narayan, D., Chambers, R., Shah, M. K. & Petesch, P. (2000). Voices of the Poor: Crying Out for Change. New York: Oxford University Press for the World Bank.

National Research Council, Division on Earth and Life Studies; Board on Earth Sciences and Resources; Committee on Private-Public Sector Collaboration to Enhance Community Disaster Resilience; Geographical Sciences Committee. (2011). Building Community Disaster Resilience Through Private-Public Collaboration. National Academies Press.

Nicolescu, B. (2002). Manifesto of Transdisciplinarity. Albany: State University of New York Press.

Nieusma, D. & Riley, D. (2010). Designs on Development: Engineering, Globalization, and Social Justice. Engineering Studies, 2(1), 29–59.

Nieusma, D. & Riley, D. (Mar. 2010). Designs on development: engineering, globalization, and social justice. Engineering Studies, 2(1).

Nieusma, D. (2004). Alternative design scholarship: Working toward appropriate design. Design Issues, 20(3), 13–24.

Nonaka, I., Von krogh, G. et al. (2006). Organizational knowledge creation theory: evolutionary paths and future advances. Organization Studies, 27(8), 1179–1208.

Normark, M., Ranstorp, M. & Ahlin, F. (2017). Financial Activities Linked to Persons from Sweden and Denmark Who Joined Terrorist Groups in Syria and Iraq During the Period 2013 – 2016, In Center for Asymmetric Threat Studies. Swedish Defence University. https://fi.se/contentassets/af6fa9 b54b8c4c6190f2cc5a8e4a0f94/terrorismfinansiering-fi-cats-2017-engny2.pdf.

Normatividad – Minería en Colombia. http://mineriaencolombia.anm.gov.co/index.php/es/normatividad (accessed Feb. 01, 2020).

NSF Award Search: Award # 1928506 – FW-HTF-RL: Collaborative Research: Shared Autonomy for the Dull, Dirty, and Dangerous: Exploring Division of Labor for Humans and Robots to Transform the Recycling Sorting Industry. (n.d.). Retrieved June 20, 2022, from https://www.nsf.gov/awardsearch/showAward?AWD_ID=1928506&HistoricalAwards=false

NSF. (n.d.) Broader Impacts https://new.nsf.gov/funding/learn/broader-impacts

Nussbaum, M. C. (2011). Creating Capabilities: The Human Development Approach. Harvard University Press.

Österberg, T., Gustafsson, B., Pedersen, P. J. & Galloway, T. A. (2015). Immigrant Child Poverty? The Achilles Heel of the Scandinavian Welfare State. In: Measurement of Poverty, Deprivation, and Economic Mobility, vol. 23, 185–219, Research on Economic Inequality 23. Emerald Group Publishing Limited.

Owens, D. A. (2011). Creative People Must Be Stopped: 6 Ways We Kill Innovation (Without Even Trying). New York, NY: Jossey-Bass.

Oxford Economics. (n.d.). The Global Green Economy: Understanding and capturing the opportunity https://www.oxfordeconomics.com/resource/the-global-green-economy-understanding-and-capturing-the-opportunity/

Pahl-Wostl, C. (2002). Participative and stakeholder based policy design, evaluation and modelling processes. Integrated Assessment, 3(1), 3–14.

Pahl-Wostl, C. (2002). Towards sustainability in the water sector – the importance of human actors and processes of social learning. Aquatic Sciences, 64(4), 394–411.

Pahl-Wostl, C. (2009). A conceptual framework for analysing adaptive capacity and multi-level learning processes in resource governance regimes. Global Environmental Change 19, 3, 354–365.

Papanek, V. J. (1971). Design for the Real World: Human Ecology and Social Change. Chicago, US: Chicago Academy.

Parkes, M. W., Bienen, L. et al. (2005). All hands on deck: transdisciplinary approaches to emerging infectious disease. EcoHealth, 2, 258–272.

Parkinson, J. (2006). Deliberating in the Real World: Problems of Legitimacy in Deliberative Democracy. Oxford: Oxford University Press.

Parsons, T. (1964). Evolutionary universals in society. American Sociological Review.

Pereira, A. G. & Funtowicz, S. (2006). Knowledge representation and mediation for transdisciplinary frameworks: tools to inform debates, dialogues & deliberations. International Journal of Transdisciplinary Research, 1(1), 34–50.

Philip, K., Irani, L. & Dourish, P. (2012). Postcolonial computing: A tactical survey. Science, Technology, & Human Values, 37(1), 3–29.

Piaget, J. (1972). The epistemology of interdisciplinary relationships. Interdisciplinarity: Problems of Teaching and Research in Universities. 127–139.

Pieth, M. (2019). Gold Laundering: The Dirty Secrets of the Gold trade – And how to clean up. Salis Verlag.

Pineda-Pinto, M., Herreros-Cantis, P., McPhearson, P. T., Frantzeskaki, N., Wang, J. & Zhou, W. Examining ecological justice within the social-ecological-technological system of New York City, USA. Landscape and Urban Planning, 215.

Pohl, C. & Hadorn, G. H. (2007). Principles for Designing Transdisciplinary Research – proposed by the Swiss Academies of Arts and Sciences. Munich, Oekom.

Polk, M. (2014). Achieving the promise of transdisciplinarity: a critical exploration of the relationship between transdisciplinary research and societal problem solving. Sustainability Science, 9(4), 439–451.

Polly Hill "The Migrant Cocoa-Farmers of Southern Ghana: A Study in Rural Capitalism" 1963 Cambridge University Press.

Polyakov, R. & Fedyanin, V. (2020). Atmospheric air pollution monitoring using flying robotics. IOP Conference Series: Materials Science and Engineering. https://doi.org/10.1088/1757-899X/1001/1/012094.

Ramadier, T. (2004). Transdisciplinarity and its challenges: the case of urban.

Rayssiguie, E. & Erden, M. S. (2022). A review of exoskeletons considering nurses. Sensors, 22(18), 7035.

Reed, M. S. (2011). Stakeholder participation for environmental management: a literature review. Biological Conservation, 141(10), 2417–2431.

"Resiliencia. Descubriendo las propias fortalezas." https://www.elpsicoanalisis.org.ar/old/numero1/resiliencia comentario1.htm (accessed Feb. 01, 2020).

Resnick, E., (Ed.). (2019). The Social Design Reader. Bloomsbury Publishing.

Restrepo, L. A. V., Patiño, L. C. A., Rueda, J. V. & Patiño, A. R. (2013). Los planes de ordenamiento territorial municipal de Antioquia. Estudio del ideario. Universidad Nacional de Colombia.

Rittel, H. W. & Webber, M. M. (1973). Dilemmas in a general theory of planning. Policy Sciences, 4(2), 155–169.

Rockström, J., Steffen, W., Noone, K., Persson, Å., Chapin, F. S., III, Lambin, E., Lenton, T. M., Scheffer, M., Folke, C., Schellnhuber, H., Nykvist, B., De Wit, C. A., Hughes, T., van der Leeuw, S., Rodhe, H., Sörlin, S., Snyder, P. K., Costanza, R., Svedin, U., Falkenmark, M., Karlberg, L., Corell, R. W., Fabry, V. J., Hansen, J., Walker, B., Liverman, D., Richardson, K., Crutzen, P. & Foley, J. (2009). Planetary boundaries: exploring the safe operating space for humanity. Ecology and Society, 14(2), 1–33.

Roco, M. C. & Bainbridge, W. S. (2002). Converging technologies for improving human performance: Integrating from the nanoscale. Journal of Nanoparticle Research, 4(4), 281–295.

Roco, M. C. & Bainbridge, W. S. (2013). The new world of discovery, invention, and innovation: convergence of knowledge, technology, and society. Journal of Nanoparticle Research, 15(9), 1–17.

Rodrik, D. (2008). The new development economics: we shall experiment, but how shall we learn?

Rohi, G., Ejofodomi, O. & Ofualagba, G. (2020). Autonomous monitoring, analysis, and countering of air pollution using environmental drones. Heliyon, 6(1). e03252. https://doi.org/10.1016/j.heliyon.2020.e03252.

Rono, J. K. (2002). The impact of the structural adjustment programmes on Kenyan society. Journal of Social Development in Africa, 17(1), 81–98.

Rosbach, D. (2012). Building a Transdisciplinary Trading Zone. The International Journal of Science in Society, 3(3), 17–30. doi: 10.18848/1836-6236/CGP/v03i03/51334.

Rosenfield, P. L. (1992). The potential of transdisciplinary research for sustaining and extending linkages between the health and social sciences. Social Science & Medicine, 35(11), 1343–1357.

Rostow, W. (1959). The Stages of Economic Growth. The Economic History Review. 12(1).

Rostow, W. W. (1959). The stages of economic growth. The Economic History Review, 12(1), 1–16.

Roy, A. (2014). The NGO-ization of Resistance. Massalijin News.

Roy, S. (2015). The Indian women's movement: Within and beyond NGOization. Journal of South Asian Development, 10(1), 96–117.

Roy, S., Adhikari, G. R. & Gupta, R. N. (Oct. 2007). Use of gold mill tailings in making bricks: a feasibility study. Waste Management & Research, 25(5), 475–482. doi: 10.1177/0734242X07076944.

RSE: Robots In Service of the Environment | Lionfish Project. (n.d.). Retrieved June 13, 2022, from https://www.robotsise.org/the-guardian-solution/

Rudin, C. (2019). Stop explaining black box machine learning models for high stakes decisions and use interpretable models instead. Nature Machine Intelligence, 1(5), 206–215.

Ruggiero, F., Lippiello, V. & Ollero, A. (2018). Aerial Manipulation: A Literature Review; Aerial Manipulation: A Literature Review. IEEE Robotics and Automation Letters, 3(3). https://doi.org/10.1109/LRA.2018.2808541.

Ruiz Córdoba, J. A., Carmona Arango, M. E. & Bolívar García, W., "Análisis del impacto ambiental generado por la emisión de mercurio desde fuentes fijas o entables del municipio de Andes – Antioquia," Instname Univ. Antioquia, 2016, Accessed: Feb. 01, 2020. [Online]. Available: http://bibliotecadigital.udea.edu.co/handle/10495/5801.

Russell, S. J. & Norvig, P. (2010). Artificial intelligence a modern approach. London.

Ryo, E. (2017). The promise of a subject-centered approach to understanding immigration noncompliance. Journal on Migration and Human Security, 5(2). 285–296. https://doi.org/10.1177/233150241700500204.

Sadowsky, M. J., Koskinen, W. C., Seebinger, J., Barber, B. L. & Kandeler, E. (2006). Automated robotic assay of phosphomonoesterase activity in soils; automated robotic assay of phosphomonoesterase activity in soils. Soil Science Society of America Journal, 70(2). 378–381. https://doi.org/10.2136/sssaj2005.0156N.

Saffari, S. (2016). Can the subaltern be heard? Knowledge production, representation, and responsibility in international development. Transcience Journal, 7(1), 36–46.

Sage, A. (2000). Transdisciplinarity: re-creating integrated knowledge, M. A. Somerville & D. J. Rapport. Oxford, UK: EOLSS Publishers Co. Ltd.

Said, E. (1978). Orientalism. New York: Pantheon.

Salzman, N., Chittoori, B., Miller, S. & Robbins, T., Senior Civil Engineering Students' Views on Sustainability and Resiliency, 2018 ASEE Annual Conference and Exposition, Jun. 2018, [Online]. Available: https://scholarworks.boisestate.edu/electrical_facpubs/395.

Samuel, A. L. (1959, July). Some studies in machine learning using the game of checkers. IBM Journal of Research and Development, 3(3), 210–229. .

Samuelson, C. C. & Litzler, E. (2016). Community cultural wealth: an assets-based approach to persistence of engineering students of color. The Journal of Engineering Education, 105(1), 93–117. doi: 10.1002/jee.20110.

Schneider, J., Lucena, J. C. & Leydens, J. A. (Winter 2009). Engineering to help: the value of critique in engineering service. IEEE Technology and Society Magazine, 28(4).

Schuler, D. & Namioka, A., (Eds.). (1993). Participatory Design: Principles and Practices. CRC Press.

Schuller, M. (2010). Unstable Foundations: Impact of NGOs on Human Rights for Port-au-Prince's Internally Displaced People. Boston, MA: Institute for Justice and Democracy in Haiti.

Scott, J. C. (1998). Seeing Like a State. London: Yale University Press.

Scott, J. C. (1998). Seeing Like a State: How Certain Schemes to Improve the Human Condition Have Failed. Yale University Press.

Scott, J. C. (2008). Seeing Like a State. Yale University Press.

Scott, P. 2017. "Data Analysis: The True Scale of Immigration and Crime in Sweden in Five Charts." The Telegraph, February 22, 2017. https://www.telegraph.co.uk/news/2017/02/22/data-analysis-true-scale-immigration-crime-sweden-five-charts/.

Selbst, A. D., Boyd, D., Friedler, S. A., Venkatasubramanian, S. & Vertesi, J. (2019, January). Fairness and abstraction in sociotechnical systems. In Proceedings of the Conference on Fairness, Accountability, and Transparency, 59–68.

Sen, A. (1982). Poverty and Famines: An Essay on Entitlement and Deprivation. Oxford University Press.

Sepulveda, A. J., Birch, J. M., Barnhart, E. P., Merkes, M., Yamahara, K. M., Marin Iii, R., Kinsey, S. M., Wright, R. & Schmidt, C. (2020). Robotic environmental DnA bio-surveillance of freshwater health. Scientific Reports, 10, 14389. https://doi.org/10.1038/s41598-020-71304-3.

Shahriari, K. & Shahriari, M. (2017, July). IEEE standard review – Ethically aligned design: A vision for prioritizing human wellbeing with artificial intelligence and autonomous systems. In: 2017 IEEE Canada International Humanitarian Technology Conference (IHTC). 197–201.

Shang, J. & Croson, R. (2009). A field experiment in charitable contribution: The impact of social information on the voluntary provision of public goods. The Economic Journal, 119(540), 1422–1439.

Shang, Jen, and Rachel Croson (2005). “Field Experiments in Charitable Contributions: The Impact of Social Influence on the Voluntary Provision of Public Goods.” 34th Annual Conference of the Association for Research on Nonprofit Associations and Voluntary Action, Washington DC, USA.

Shang, Jen and Rachel Croson, 2009. A field experiment in charitable contribution: The impact of social information on the voluntary provision of public goods. *The Economic Journal* 119(540): 1422–1439.

Sharkey, N. & Sharkey, A. (2012). The eldercare factory. Gerontology, 58(3). 282–288, https://doi.org/10.1159/000329483.

Sheller, M. (2020). Island Futures: Caribbean Survival in the Anthropocene. Durham: Duke University Press.

Sheller, M., Galada, H. C., Montalto, F., Gurian, P. L., Piasecki, M., Ayalew, T. B. & O’Connor, S. (2013). Gender, disaster and resilience: assessing women’s water and sanitation needs in Léogâne, Haiti, before and after the 2010 earthquake. wH2O: The Journal of Gender and Water, 2(1), 18–27.

Sheller, M., O’Connor, S., Galada, H. C., Montalto, F. A., Gurian, P. L. & Piasecki, M. (2015). Participatory Engineering for Recovery in Post-Earthquake Haiti. Engineering Studies, 6(3), 159–190. SI on Engineering Risk and Disaster.

Short, J., Jr. (2018). Poverty, Ethnicity, And Violent Crime. Routledge.

Shukla, A. & Karki, H. (2016). Application of robotics in onshore oil and gas industry – a review part I. Robotics and Autonomous Systems, 75, 490–507. https://doi.org/10.1016/j.robot.2015.09.012.

Siegel, S. & Veiga, M. M. (Jan. 2010). The myth of alternative livelihoods: artisanal mining, gold and poverty. The International Journal of Environment and Pollution, 41(3–4), 272–288. doi: 10.1504/IJEP.2010.033236.

Silberman, N., Ahrlich, K., Fergus, R. & Subramanian, L. (2010, March). Case for automated detection of diabetic retinopathy. In: 2010 AAAI spring symposium series.

Simon, H. (1969). The Sciences of the Artificial. MIT Press.

Sims, J., Federal Reserve Bank of Chicago, Romero, J., & Federal Reserve Bank of Richmond. (2013). Latin American Debt Crisis of the 1980s.

Sinai, N., Leftwich, D., & McGuire, B. (2020). Human-centered policymaking: What government policymaking can learn from human-centered design and agile software development. Belfer Center for Science and International Affairs, Harvard Kennedy School. https://www.belfercenter.org/publication/human-centered-policymaking.

Slingerland, E. (2000). Effortless action: The Chinese spiritual ideal of wu-wei. Journal of the American Academy of Religion, 68(2), 293–327.

Smith, A. (1776). The Wealth of Nations. London, England: W. Strahan and T. Cadell.

Smith, J. & Lucena, J. C. Socially Responsible Engineering. In: Routledge Handbook of Philosophy of Engineering. London: Taylor & Francis.

Smith, N. (1984). Uneven Development: Nature, Capital, and the Production of Space. Oxford: Basil Blackwell.

Smith, V. L. (1962). An experimental study of competitive market behavior. Journal of Political Economy, 70, 111–137.

Smith, V. L. (1964). The effect of market organization on competitive equilibrium. Quarterly Journal of Economics, 78, 181–201.

Smith, V. L. (1976). Experimental economics: Induced value theory. American Economic Review, 66(2), 274–279.

Smith, V. L. (1982). Microeconomic systems as an experimental science. American Economic Review, 72(5), 923–955.

Smith, V. L., Suchanek, G. L. & Williams, A. W. (1988). Bubbles, crashes, and endogenous expectations in experimental spot asset markets. Econometrica, 56, 1119–1151.

Smits, K. M. et al. (Dec. 2019). Voces Mineras: Clarifying the future of artisanal and small-scale mining collaborations. The Extractive Industries and Society. doi: 10.1016/j.exis.2019.12.003.

Soininen, M. (1999). The ‘Swedish Model’ as an Institutional Framework for Immigrant Membership Rights. Journal of Ethnic and Migration Studies, 25(4), 685–702.

Somerville, M. A. & Rapport, D. J. (Eds.). (2000). Transdisciplinarity: re-creating integrated knowledge. Oxford, UK: EOLSS Publishers Co. Ltd.

Sparrow, R. (2020). Robotics has a race problem. Science, Technology, and Human Values, 45(3), 538–560.

Steffen, W., Richardson, K., Rockström, J., S. E. Cornell, J., Fetzer, I., Bennett, E. M., Biggs, R., Carpenter, S. R., De vries, W., de Wit, C. A., Folke, C., Gerten, D., Heinke, J., Mace, G. M., Persson, L. M., Ramanahthan, V., Reyers, B. & Sörlin, S. (2015). Planetary boundaries: Guiding human development on a changing planet. Science, 347(6223), 1259855.

Stephan, P. F. (2015). Designing 'matters of concern'(Latour): A future design challenge. In: W. Jonas, S. Zerwas & K. Von anshelm (Eds.), Transformation design. 202–226.

Stettler Kleine, M., Zacharias, K. & Ozkan, D. S. (2021, July), Contextualization as Virtue in Engineering Education. Paper presented at 2021 ASEE Virtual Annual Conference Content Access, Virtual Conference. doi: 10.18260/1-2-36844

Stiglitz, J. E. (2002). Participation and development: Perspectives from the comprehensive development paradigm. Review of Development Economics, 6(2), 163–182.

Stokols, D. (2006). Toward a Science of Transdisciplinary Action Research. American Journal of Community Psychology, 38, 63–77.

Stokols, D., Fuqua, J. et al. (2003). Evaluating transdisciplinary science. Nicotine & Tobacco Research, 5(1), S21–S39.

Stokols, D., Hall, K. L. et al. (2008). The science of team science: overview of the field and introduction to the supplement. American Journal of Preventive Medicine, 35(2S), S77–S89.

Stokols, D., Misra, S. et al. (2008). The ecology of team science: understanding the contextual influences on transdisciplinary collaboration. American Journal of Preventive Medicine, 35(2S), S96–S115.

Stretmo, L. & Hammarén, N. (2022). Dangerous Youngsters or Youngsters (Not) in Danger? Constructions of Unaccompanied Children by Swedish Service Providers. Journal of Ethnic and Migration Studies, 1–18. https://doi.org/10.1080/1369183X.2022.2053079.

Struckmann, C. (Jan. 2018). A postcolonial feminist critique of the 2030 Agenda for Sustainable Development: A South African application. Agenda, 32(1), 12–24. doi: 10.1080/10130950.2018.1433362.

Stuart, H., Wang, S., Khatib, O. & Cutkosky, M. R. (2017). The Ocean One hands: An adaptive design for robust marine manipulation. 36(2), 150–166. https://doi.org/10.1177/0278364917694723.

Sutton, R. I. & Hargadon, A. (1996). Brainstorming groups in context: Effectiveness in a product design firm. Administrative Science Quarterly, 41(4), 685–718.

Swyngedouw, E. (2004). Social Power and the Urbanization of Water: Flows of Power. Oxford: Oxford University Press.

Swyngedouw, E. (2007). The post-political city. Urban Politics Now, 58–76.

Sze, J. & London, J. K. (2008). Environmental justice at the crossroads. Sociology Compass, 2(4), 1331–1354.

Szegedy, C., Zaremba, W., Sutskever, I., Bruna, J., Erhan, D., Goodfellow, I., & Fergus, R. (2013, December). Intriguing properties of neural networks. In 2nd International Conference on Learning Representations (ICLR) 2014.

Tai, E. T. T. (2018). Liability for (semi) autonomous systems: robots and algorithms. In: Research Handbook in Data Science and Law. Edward Elgar Publishing, 55–82.

Telliel, Y. & Krueger, R. (2020). Decentering Design Thinking for Development Engineering. In: Proceedings of International Conference on Design History and Design Studies. Zagreb, Croatia.

Telliel, Y. D. & Kruegar, R. (2020). Decentring Design Thinking for Development Engineering, In: Lessons to Learn? Past Design Experiences and Contemporary Design Practices. https://www.academia.edu/44758306/Decentring_Design_Thinking_for_Development_Engineering.

Thomas, H. & Santos, G. (2016). (Sociologist), Tecnologías para incluir: ocho análisis socio-técnicos orientados al diseño estratégico de artefactos y normativas. Instituto de Estudios sobre la Ciencia y la Tecnología, IESCT, Universidad Nacional de Quilmes.

Thompson, J. & Doherty, B. (2006). The diverse world of social enterprise: A collection of social enterprise stories. International Journal of Social Economics, 33(5–6), 361–375.

Tischler, J. (2013). Light and power for a multiracial nation: The Kariba Dam scheme in the Central African Federation. Springer.

Tischler, Julia. 2013. *Light and power for a multiracial nation: The Kariba Dam scheme in the Central African Federation*. Springer.

Tiwari, K. & Chong, N. Y. (2019). Multi-robot exploration for environmental monitoring: The resource constrained perspective, In: Multi-robot Exploration for Environmental Monitoring: The Resource Constrained Perspective. Elsevier. https://doi.org/10.1016/C2018-0-01886-1.

Tomašev, N., Cornebise, J., Hutter, F., Mohamed, S., Picciariello, A., Connelly, B. . . . Clopath, C. (2020). AI for social good: unlocking the opportunity for positive impact. Nature Communications, 11(1), 2468.

Tuler, S., Dow, K. & Webler, T. (2020). Assessment of adaptation, policy, and capacity building outcomes from 14 processes. Environmental Science and Policy, 114, 275–282.

Tuler, S., Dow, K., Webler, T. & Whitehead, J. (2016). Learning through participatory modeling: Reflections on what it means and how it is measured. In: S. Grey, M. Paolisso, R. Jordan & S. Grey (Eds.), Including Stakeholders in Environmental Modeling. Switzerland: Springer International Publishing.

Tuomas, J., Lukka, T., Tossavainen, J. V., Kujala, D. & Raiko, T. (2014). ZenRobotics Recycler-Robotic Sorting using Machine Learning.

UNICEF. (2007). An Evaluation of the PlayPump® Water System as an Appropriate Technology for Water, Sanitation and Hygiene Programmes. UNICEF, https://www-tc.pbs.org/frontlineworld/stories/southernafrica904/flash/pdf/unicef_pp_report.pdf.

United Nations. Department of Economic Affairs. (1951). Measures for the Economic Development of Under-developed Countries: Report by a Group of Experts Appointed by the Secretary-General of the United Nations (Vol. 51). United Nations, Department of Economic Affairs.

US AID. (2019) Digital Development Strategy 2020–2024, accessed at https://www.usaid.gov/sites/default/files/2022-05/USAID_Digital_Strategy.pdf.pdf.

Vakilian, K. A. & Massah, J. (2017). A farmer-assistant robot for nitrogen fertilizing management of greenhouse crops. Computers and Electronics in Agriculture, 139, 153–163. https://doi.org/10.1016/j.compag.2017.05.012.

Veiga, M. M. & Marshall, B. G. (Jan. 2019). The Colombian artisanal mining sector: Formalization is a heavy burden. The Extractive Industries and Society, 6(1), 223–228. doi: 10.1016/j.exis.2018.11.001.

Vinuesa, R. & Sirmacek, B. (2021). Interpretable deep-learning models to help achieve the Sustainable Development Goals. Nature Machine Intelligence, 3(11), 926–926.

VisDA2022: Visual Domain Adaptation Challenge. (n.d.). Retrieved June 20, 2022, from https://ai.bu.edu/visda-2022/

Voinov, A. & Bousquet, F. (2010). Modeling with stakeholders. Environmental Modelling and Software, 25(11), 1268–1281.

Voinov, A. & Brown Gaddis, E. J. (2008). Lessons for successful participatory watershed modeling: a perspective from modeling practitioners. Ecological Modelling, 216(2), 197–207.

Wade, R. H. (2016). Boulevard to broken dreams, Part 1: the Polonoroeste road project in the Brazilian Amazon, and the World Bank's environmental and indigenous peoples' norms. Brazilian Journal of Political Economy, 36(1), 214–230.

Wagenaar, H. (2011). Meaning in Action: Interpretation and Dialogue in Policy Analysis. Armonk, NY: M.E. Sharpe.

Wampler, B. (2000). A guide to participatory budgeting. World Bank.

Wangui, K. A. (2021). *Internal organization environment and project performance in Nairobi City County projects* (Doctoral dissertation, Kenyatta University).

Ward, A. S. & Bruce, A. (2022). The Tech That Comes Next: How Changemakers, Philanthropists, and Technologists Can Build an Equitable World. John Wiley & Sons.

Warford, E. (2018, June), Toward a More Caring Code of Engineering Ethics. Paper presented at 2018 ASEE Annual Conference & Exposition, Salt Lake City, Utah.

Warner, J. T. & Pleeter, S. (2001). The personal discount rate: Evidence from military downsizing programs. American Economic Review, 91(1), 33–53.

Waste Robotics. (n.d.). Retrieved February 25, 2019, from https://wasterobotic.com/

Watters, L. (2018). How technology is changing the global development jobs landscape. https://www.devex.com/news/how-technology-is-changing-the-global-development-jobs-landscape-92970, Retrieved 06/15/2023.

Weick, K. E. (1995). Sensemaking in Organizations. Thousand Oaks, CA: Sage.

White, D. (2020). Just transitions/design for transitions: preliminary notes on a design politics for a green new deal. Capitalism Nature Socialism, 31(2), 20–39.

White, D. (2020). Just transitions/design for transitions: preliminary notes on a design politics for a green new deal. Capitalism Nature Socialism, 31(2), 20–39.

Whiteley, N. (1993). Design for Society. London, UK: Reaktion books.

Wiesbrock, A. (2011). The Integration of Immigrants in Sweden: A Model for the European Union? International Migration, 49(4), 48–66.

Williamson, J. G. (2006). Poverty Traps, Distance, and Diversity: The Migration Connection. 12549.

Winschiers-Theophilus, H., Bidwell, N. J. & Blake, E. (2012). Community consensus: design beyond participation. Design Issues, 28(3), 89–100.

Wong, S. (2012). What have been the impacts of World Bank Community-Driven Development Programs? CDD Impact Evaluation Review and Operational and Research Implications. World Bank, Washington, DC, 7.

Wood, T. & Murray, W. E. (2007). Participatory democracy in Brazil and local geographies: Porto Alegre and Belo Horizonte compared. Revista Europea de Estudios Latinoamericanos y del Caribe/European Review of Latin American and Caribbean Studies, 19–41.

Wrzesniewski, A. (2003). Finding positive meaning in work. Positive Organizational Scholarship: Foundations of a New Discipline, 308.

Yeh, C., Perez, A., Driscoll, A., Azzari, G., Tang, Z., Lobell, D., Ermon, S., & Burke, M. (2020, May). Using publicly available satellite imagery and deep learning to understand economic well-being in Africa. Nature Communications.

Yuval-Davis, N. (2006). Belonging and the Politics of Belonging. Patterns of Prejudice, 40(3), 197–214.

Zahugi, E. M. H., Shanta, M. M. & Prasad, T. V. (2013). Oil spill cleaning up using swarm of robots. Advances in Intelligent Systems and Computing, 178, 215–224. https://doi.org/10.1007/978-3-642-31600-5_22/COVER/.

Zaoui, M., Viguier, G. & Blazevic, P. (2015). Mobile Robotics for Restoring Degraded Ecosystems. IEEE 2015 Global Humanitarian Technology Conference.

Zeldovich, L. (2019). Underwater drone hunts coral-eating crown-of-thorns starfish. American Society of Mechanical Engineers. https://www.asme.org/topics-resources/content/underwater-drone-hunts-coraleating-crownofthorns.

Zhou, M., Lee, J., Vallejo, J. A., Tafoya-Estrada, R. & Xiong, Y. S. (2008). Success Attained, Deterred, and Denied: Divergent Pathways to Social Mobility in Los Angeles's New Second Generation. The Annals of the American Academy of Political and Social Science, 620(1). 37–61. https://doi.org/10.1177/0002716208322586.

Zhou, Y., Dong, F., Kong, D. & Liu, Y. (2019). Unfolding the convergence process of scientific knowledge for the early identification of emerging technologies. Technological Forecasting and Social Change, 144, 205–220.

Zierhofer, W. & Burger, P. (2007). Disentangling transdisciplinarity: an analysis of knowledge integration in problem-oriented research. Science Studies, 20(1), 51–74.

Zierhofer, W. & Burger, P. (2007). Disentangling transdisciplinarity: an analysis of knowledge integration in problem-oriented research. Science & Technology Studies, 20(1), 51–72.

Index

https://doi.org/10.1515/9783112218716-017

Integrated Global STEM Series

Volume 1
Science, Engineering, and Sustainable Development: Cases in Planning, Health, Agriculture, and the Environment
Robert Krueger, Yunus Telliel, Wole Soboyejo (Eds.), 2024
ISBN: 978-3-11-075749-1, e-ISBN: 978-3-11-075750-7

Volume 2
Smart Villages: Generative Innovation for Livelihood Development
Robert Krueger, Hermine Vedogbeton, Mustapha Fofana, Wole Soboyejo (Eds.), 2024
ISBN: 978-3-11-078621-7, e-ISBN: 978-3-11-078623-1

Volume 3
Integrated Agriculture: An Approach for Sustainable Agriculture
Amanullah, 2024
ISBN: 978-3-11-144801-5, e-ISBN: 978-3-11-144807-7

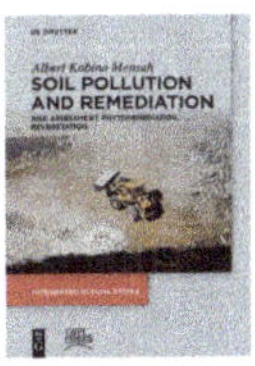

Volume 4
Soil Pollution and Remediation: Risk Assessment, Phytoremediation, Revegetation
Albert Kobina Mensah, 2025
ISBN: 978-3-11-166038-7, e-ISBN: 978-3-11-166204-6

Volume 5
Emerging Technologies in Africa
Matthew N. O. Sadiku, 2025
ISBN: 978-3-11-914583-1, e-ISBN: 978-3-11-221198-4

Volume 6
Decentering Science and Technology in Global Development: Co-Designing for Community Sustainability
Robert Krueger, 2025
ISBN: 978-3-11-914352-3, e-ISBN: 978-3-11-221871-6

www.ingramcontent.com/pod-product-compliance
Lightning Source LLC
Chambersburg PA
CBHW061210240926
51846CB00006B/1171
* 9 7 8 3 1 1 9 1 4 3 5 2 3 *